D1453510

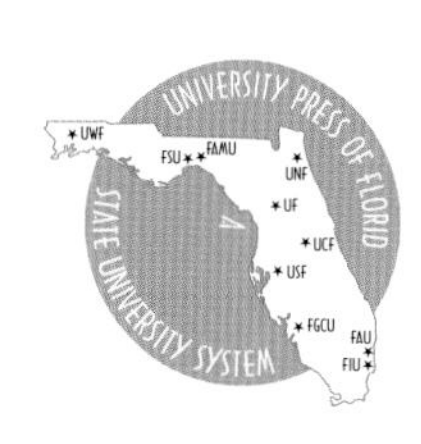

Florida A&M University, Tallahassee
Florida Atlantic University, Boca Raton
Florida Gulf Coast University, Ft. Myers
Florida International University, Miami
Florida State University, Tallahassee
University of Central Florida, Orlando
University of Florida, Gainesville
University of North Florida, Jacksonville
University of South Florida, Tampa
University of West Florida, Pensacola

University Press of Florida
Gainesville · Tallahassee
Tampa · Boca Raton
Pensacola · Orlando
Miami · Jacksonville
Ft. Myers

Part 1

Using the Guide

An Introduction to Orchids

Orchids hold a special fascination for many people, perhaps because of their perceived extreme beauty, rarity, and mystery. In actuality, orchids are the largest family of flowering plants on earth with nearly 30,000 species. Although many are exceedingly beautiful, many more are small and dull colored, barely 2 or 3 mm across! There is hardly any place on this planet, other than the Antarctic, that does not have some species of orchid growing within its native flora. Even the oases within the great deserts of the world harbor a few species. In the more northerly climes orchids can be found well above the Arctic Circle. They grow at elevations above 15,000 feet, and within the highly developed urban areas of the globe there are still orchids persisting.

In the southeastern United States orchids abound in all nine states. Some of the rarest species globally are to be found in this area. Because of their geographic position in North America, the southeastern states have become home to the southernmost locales for several northern species, and the northernmost locales for many southern species. In addition several species are found primarily here, and one, *Platanthera chapmanii,* is endemic.

Because of the sheer number of orchids in the world there is an enormous amount of morphological diversity. Although orchids all possess a certain number of qualifying characters, their general morphology can be as variable as the imagination. But, viewed closely under a lens, even the tiniest of orchids has the distinctive characters that make it an orchid!

Characteristics of the family Orchidaceae are quite simple, despite their diversity. First, they are monocotyledons, or monocots—a major class of the plant kingdom that has a single emerging leaflike structure when the seed germinates (as opposed to dicotyledons, or dicots, which have two leaflike structures). Grasses, lilies, and palms are also monocots. Second, orchids have three sepals, two petals, and a third petal that is modified into a lip. This prominent structure is actually a guide for the pollinator—a kind of landing platform that guides the agent of pollination toward the nectary, where it passes by the column and, in doing so, effects fertilization. Again, because of the size of the family, there are many pollinators, not just insects. Orchids have been documented to be pollinated by the usual bees, butterflies, wasps, and flies, as well as by hummingbirds, and a few are even rain-assisted in their pollination. Many orchids emit a fragrance at a certain time of day to attract their specific pollinator. Beyond that there are those species that by various means are self-pollinating. Third, the stamens and pistil are united into a column, a structure unique to the orchids.

Because of the number of genera and species, occasionally there are what appear to be exceptions. A few genera have monoecious or dioecious flowers—male and female flowers on separate inflorescences or plants. In others, the petals and lip may be so modified as to be barely recognizable, and, in yet others, the flowers never open: they are cleistogamous and are self-fertilized in the bud. But all of these exceptions aside, the vast majority of orchid flowers do look like orchids!

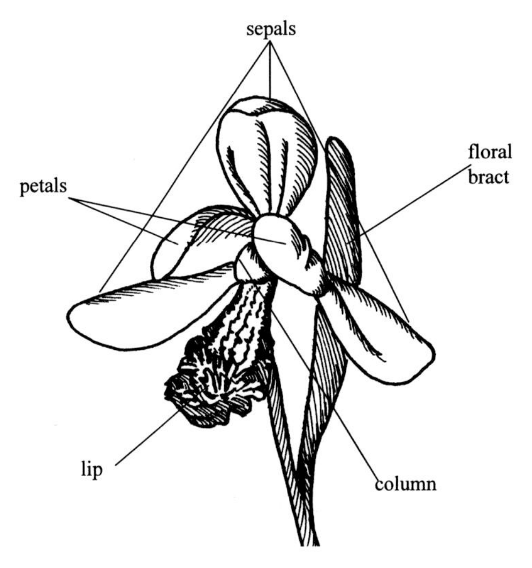

In the southeastern United States, as defined in this field guide, we have both terrestrial and epiphytic orchids. The terrestrials grow with the roots within the ground and take in water and nutrients through root hairs and/or swollen stems that are tuber-, bulb-, or corm-like. Epiphytic orchids live within the trees, their roots covered with a hard coating called a velamen layer, and the tips are often green and soft so as to absorb water and nutrients. Although many epiphytes found elsewhere have pseudobulbs, swollen stems that act as water storage organs, the only epiphyte to be found within this range is *Epidendrum magnoliae*, the **green-fly orchis**, which has simple canelike stems. Although not typical of any of the orchids of the southeastern United States, there are also some species that are lithophytic—growing on rock strata.

Wild Orchids

of the Southeastern United States, North of Peninsular Florida

Paul Martin Brown

with Drawings by Stan Folsom

Printed in China on acid-free paper

09 08 07 06 05 04 6 5 4 3 2 1

A record of cataloging-in-publication data is available from the Library of Congress.
ISBN 0-8130-2748-9 (cloth)
ISBN 0-8130-2749-7 (paper)

The University Press of Florida is the scholarly publishing agency for the State University System of Florida, comprising Florida A&M University, Florida Atlantic University, Florida Gulf Coast University, Florida International University, Florida State University, University of Central Florida, University of Florida, University of North Florida, University of South Florida, and University of West Florida.

University Press of Florida
15 Northwest 15th Street
Gainesville, FL 32611-2079
http://www.upf.com

Alexander J. Sneewitchen

In Memoriam

1984–2003

the best catpanion any botanist ever had

BUSINESS/SCIENCE/TECHNOLOGY DIVISION
CHICAGO PUBLIC LIBRARY
400 SOUTH STATE STREET
CHICAGO, ILLINOIS 60605

A reminder

Our wild orchids are a precious resource. For that reason they should never be collected from their native habitats, either for ornament or home gardens. All orchids grow in association with specific fungi and these fungi are rarely present out of the orchids' original home. Searching for and finding many of these choice botanical treasures is one of the greatest pleasures for both the professional and amateur botanist. Please leave them for others to enjoy as well.

Contents

Foreword

Man has had a fascination for orchids dating back to Confucius, who wrote of the Chinese interest in and use of *lan* (Chinese for orchid) right on down through the ages. Unfortunately, man has also been one of the most destructive elements in the disappearance of orchids. As populations grow native habitats succumb to subdivisions or agricultural and lumbering enterprises. Hence, it is books like Paul Martin Brown's *Wild Orchids of the Southeastern United States* that play an important role in making us aware of our native orchid flora so that we can be proactive in assuring that these orchids will be here for generations to come.

There have been many books written about our native orchids, such as W. H. Gibson's *Our Native Orchids* (1905) and *Native Orchids of North America* by D. Correll (1950), a very comprehensive volume that was the bible for many years. These were followed by Carlyle Luer's *The Native Orchids of Florida* (1972) and *The Native Orchids of the United States and Canada excluding Florida* (1975). During the time since these comprehensive works were published new species have been discovered and many nomenclatural changes have taken place. Paul Martin Brown brings orchidologists up to date in *Wild Orchids of the Southeastern United States.*

Paul Martin Brown is no newcomer to the world of orchidology. He has a Masters degree from the University of Massachusetts, has taught for a number of universities and organizations, and is presently a research associate at the University of Florida. He founded the North American Native Orchid Alliance in 1995 and is also the founding and present editor of the *North American Native Orchid Journal.* In addition, Paul and his partner Stan Folsom have published, since 1997, *Wild Orchids of the Northeastern United States* (1997), *Wild Orchids of Florida* (2002), and *The Wild Orchids of North America, North of Mexico* (2003).

This latest work, *Wild Orchids of the Southeastern United States*, although intended as a field guide, is much more comprehensive than that. This thorough book is divided into four parts. The first section introduces us to orchids, explains what they are, and includes keys to help identify them. The second part contains complete descriptions, color illustrations, and black and white drawings, the latter by Stan Folsom, for all the known species native to the area. Part three contains a variety of resources, including a checklist, synonyms, and references. Part four divides the southeastern United States into eight geographic localities, lists all the species inhabiting each area, and provides descriptions of the types of growing conditions found in each area. In addition, the Latin and common names for the species are included, making identification easier for general users. Paul Martin Brown has done an exhaustive job not only in enumerating all the species, including recently discovered species, but also in bringing the nomenclature up to date, making this volume a valuable tool for years to come.

Tom Sheehan

Preface

For the purpose of this field guide, the southeastern United States is defined in the map below. It includes the eastern counties of Texas, southeastern Arkansas, all of Louisiana, Mississippi, Alabama, Georgia, South Carolina, southeastern North Carolina, and the panhandle and northeastern counties of Florida. Many species typical of the Gulf and Atlantic Coastal Plains extend well into these areas, while the southern Appalachian Mountains of northern Alabama, Georgia, and South Carolina have species at the southern limits of their more northerly ranges.

Although no individual work treats this entire region, portions have been treated, in various ways, in a variety of publications. *Wild Orchids of Texas* (Liggio and Liggio, 1999), *Wild Orchids of Arkansas* (Slaughter, 1995), *Wild Orchids of South Carolina* (Fowler, forthcoming 2005), and *Wild Orchids of Florida* (Brown and Folsom, 2002), cover those states in their entirety. Private individuals and state agencies have often developed county distribution maps for many of the states and individual genera have been treated in some cases. In all genera found

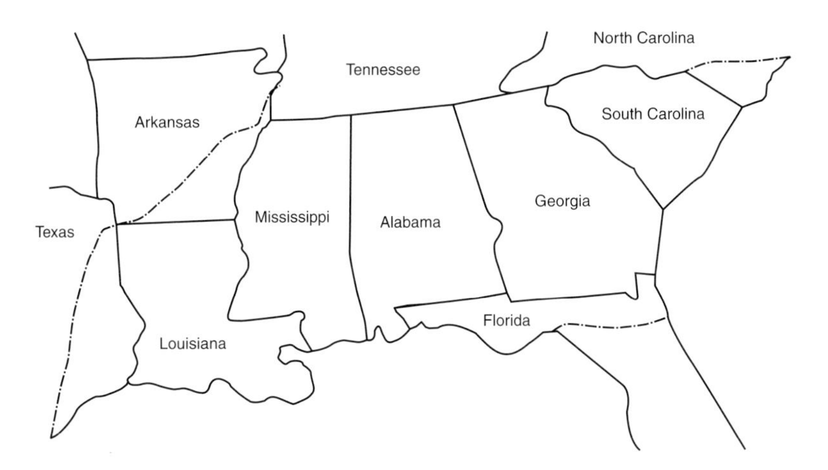

in the Southeast, the recent work in volume 26 of *Flora of North America* (2002) has greatly fine-tuned the distribution of the orchids found in this region. The genus that presents the greatest problem has been, and still is, *Spiranthes*.

The descriptions of *Spiranthes eatonii* (1999) and *S. sylvatica* (2002) have gone a long way in clearing up some of the confusion among the spring-flowering species, but the presence of several races of *S. cernua*, including the recently rediscovered Deep South race, still leads to frequent misidentifications of specimens. These problems are compounded by the perpetuation of erroneous material in both earlier and recent literature. These problematic species are resolved within the individual species treatments.

A Note About Field Guides

The primary purpose of a field guide is to assist the user in identifying, in this case, the wild orchids of the southeastern United States. It is not intended to be either an exhaustive treatise on those species, nor the ultimate reference for that area. It is intended to be used primarily in the field and is designed for locating information easily while one foot is in the proverbial bog. The photographs have been taken in the field and are intended to illustrate the species as the user will see them. They are neither studio shots, nor great works of art. Just good diagnostic photos that portray the plants in their habitats. The line drawings also have the same goal, that of assisting the user in identifying the orchids. Many times line drawings can communicate different aspects of the plants that photographs cannot. The user is always encouraged to consult additional references for more detailed descriptions and regional accounts for individual species. It is always possible to write more on any subject and more certainly could be said for each species, but first and foremost this is a field guide. By combining the resources of the keys, descriptions, photographs, and line drawings the user should be able, with relative simplicity, to identify all of the orchids within the range of this book.

Acknowledgments

This work, with such a large geographic scope, has involved the advice and counsel of many people. Those who facilitated the research and made major contributions to the information gathered within this book include Norris Williams, keeper, and Kent Perkins, manager, of the University of Florida Herbarium, Florida Museum of Natural History, Gainesville; J. Richard Carter, Valdosta State University, Valdosta, Georgia; John B. Nelson, Moore Herbarium, University of South Carolina, Columbia; Richard D. Porcher, The Citadel, Charleston; Albert E. Sanders, Charleston Museum, South Carolina; Gustavo Romero, Orchid Herbarium of Oakes Ames, Judith Warnement, Botany Libraries, Harvard University Herbaria; Scott Stewart, Ann Malmquist, Carl R. Slaughter, M.D. (Arkansas); Stan Bentley (North Carolina); Joe Liggio, Barney Lipscomb (Texas); Gary Landry (Louisiana); Tom Patrick, Jim Allison (Georgia); Jim Fowler (South Carolina); Loran Anderson, Linda Chafin, Gil Nelson, Guy Anglin, Pam and Bill Anderson, Joel De Angelis (Florida); Al Schotz, Caroline Dean (Alabama); Ron Wieland (Mississippi).

Ken Scott and his staff at University Press of Florida have continued their usual support, without which this book would never have been completed. And, again, my partner, Stan Folsom, whose paintings and drawings grace the pages of this book, has been a source of support and inspiration through the entire project. For that I am most grateful.

Abbreviations and Symbols

ca. = about or approximate, in reference to measurements

cm = centimeter

f. = *filius*; son of, or the younger

m = meter

mm = millimeter

nm = nothomorph or nothovariety, indicating an intraspecific hybrid

subsp. = subspecies

var. = variety

× between or preceding a name denotes a hybrid or hybrid combination

* = naturalized

≠ = misapplied name

Publications

AOS = American Orchid Society

FNA = *Flora of North America* (volume 26, including the *Orchidaceae*)

NANOJ = *North American Native Orchid Journal*

In nature, all orchids consume fungi as a food (carbon) source. Each wild orchid has this unique relationship with naturally occurring fungi. While nearly impossible to see with the naked eye, these fungi are essential to the growth and development of any wild orchid. In this relationship, a fungus infects the orchid roots and the fungus is then consumed as a food source, initially prompting seed germination and sustaining seedling development. As mature plants, orchids are capable of producing food via both photosynthesis and the consuming of its root fungus. A number of different fungi may contribute nutrients, growth regulators, vitamins, and moisture at various levels and at different times throughout the orchid's life cycle. It is this special relationship between orchid and fungus that makes native orchids extremely difficult to transplant. In most cases, native orchids that have been moved from their natural location will die within a few years from this disruption of the orchid-fungal association.

The leaves of orchids are nearly as diverse as the flowers. As is characteristic of all monocots, they have parallel primary veins. The leaves may be long and slender, grasslike, round and fat, hard and leathery, or soft and hairy—just about any configuration. If you are a novice, before you start to use this book take a few moments to look through the photographs and drawings and see the typical floral parts. Try to get a better feeling for what an orchid is and then go out and find them!

Which Orchid Is It?

A key is a written means for identifying an unknown species. The principles used in keying are very simple. A series of questions are asked, and the user needs to determine a positive or negative response. Couplets are used to ask these questions. Be sure to read both halves of the couplet before deciding on your answer. The correct answer in the couplet that matches your observation of the orchid directs you to the next set of couplets. If you follow them, correctly matching each observation, you eventually key out the genus and/or species. As you progress through the key, the illustrations may help you in making your determinations.

Choose carefully the specimen you wish to identify. Look for a typical, average plant—neither the largest nor the smallest. *This key is designed so you should be able to use it without picking any of the orchids.* Only in the case of a few similar species will detailed examinations be necessary. The use of measurements has been kept to a minimum, as has the use of color. Be aware that white-flowered forms exist in many of the species and they usually occur with the typical color form.

Before you start to use the key you should always mentally note the following:

1. placement and quantity of leaves, i.e., basal vs. cauline (see glossary); opposite vs. alternate; 1, 2, or more
2. placement and quantity of flowers, i.e., terminal vs. axillary; single vs. multiple
3. geographic location and habitat

Six vocabulary words that will help in your understanding of the key:

terrestrial—growing with the roots in the ground
epiphytic—growing with the roots exposed in the air, usually on tree bark
lithophytic—growing on rocks
pseudobulb—the swollen storage organ at the base of the leaves, primarily on epiphytes, occasionally on terrestrials

bract—a small, reduced leaf that usually is found on the flowering stem and/or within the inflorescence

spur/mentum—a projection at the base of the lip; it may be variously shaped from slender and pointed (spur) to short and rounded (mentum)

Other terms are illustrated within the key or can be found in the glossary on page 363.

Using the Key

If you are using the key for the first time, start with a species with which you are familiar—perhaps the **common grass-pink**, *Calopogon tuberosus*.
Starting with couplet 1

1a plants epiphytic, or rarely lithophytic . . . *Epidendrum*, p. 65
1b plants terrestrial . . . 2
which takes us to couplet 2

2a lip inflated or sac shaped . . . 3
2b lip otherwise . . . 5
which takes us to couplet 5

5a (green) leaves (apparently) lacking at flowering time; stem bracts may be present . . . 6
5b (green) leaves present at flowering time . . . 12
which takes us to couplet 12

12a spur or mentum present . . . 13
12b spur or mentum lacking . . . 20
which takes us to couplet 20

20a pseudobulbs present, although they may be well hidden in the leaf bases . . . 21
20b pseudobulbs lacking . . . 22
which takes us to couplet 22

22a leaves essentially basal or extending up the lower 1/4 of the stem and rapidly reduced to leafy bracts . . . 23
22b leaves essentially cauline . . . 25
which takes us to couplet 25

25a flowers non-resupinate . . . *Calopogon*, p. 25
25b flowers resupinate . . . 26

To continue to species, use the species key for *Calopogon* on page 25 and follow the same procedure.

This key is constructed for use in the field or with live specimens, and is based on characters that are readily seen. It is not a technical key in the strictest sense, but simply intended to aid in field identification.

Keys are not difficult if you take your time, learn the vocabulary, and hone your observational skills. Like any skills, the more you use them the easier it becomes.

Some Important Notes About . . .

Plant Names

Examples: *Platanthera integrilabia* (Correll) Luer
Cleistes bifaria (Fernald) Catling & Gregg

The Latin name used is composed of a genus and a species. The genus (plural, genera) is the broader group to which the plant belongs and the species is the specific plant being treated. After the two Latin names, the name or names of the people who first described the plant appear. The two examples above illustrate two different author citations. In the first example Correll originally described the species as *Habenaria integrilabia,* therefore his name appears immediately after the genus and species. Subsequently Luer transferred it to the genus *Platanthera,* therefore his name appears after Correll's, which has been placed in parentheses to indicate that Correll was the original describer. In the second example Fernald was the first person to describe this taxon, although he named it at the varietal level, *Cleistes divaricata* var. *bifaria;* therefore his name comes first. Catling & Gregg validated the variety as a species and therefore their names follow. Other ranks may occur, such as subspecies, variety, and forma. Subspecies and variety usually designate a variation that has a significant difference from the species and a definite geographic range as in *Platanthera flava* var. *herbiola,* hence it is a variety. Varieties and subspecies should always breed true. There can always be a bit of disagreement over variety and subspecies and which term is better used. Color variations, which occur throughout the range of the species, are best treated as forma, as in *Platanthera grandiflora* forma *albiflora* (Rand & Redfield) Catling—white-flowered form. These forms can, and do, occur randomly and rarely breed true. All forma that have received names for the species occurring within the southeastern United States are listed, although many have yet to be documented for this region.

Common Names

Common names are never as consistent throughout the range of the plants as we would like. The most frequently used common names are listed in the plant descriptions, as well as some regional names.

Orchid or *orchis*? For common names either *orchid* or *orchis* may be used, but traditionally *orchis* has been used for certain genera and that has been maintained here.

Size

Average height of the plants, floral size, and number of flowers are given, with extremes in parentheses. The relative scale of each line drawing is for an average plant and is based on the flower size.

Color

The color of the flowers as it normally occurs is given. Remember to check to see if color variants such as white-flowered forms occur. In some genera the overall color of the petals and sepals, the perianth, is given and the lip color follows.

Forms and Hybrids

The diversity of forms, randomly occurring variations such as color and habit, and hybrids can take hunting for native orchids to another level. After mastering many of the species one can be challenged to search for all the variations. It would be unrealistic to think that all of these forms occur within the range of this work, but many of them do. To better understand these forms, note that each is annotated with the original publication and type locality of the taxon. Those that have been documented within the Southeast are noted within the text and it is expected that many of the other forms will also be found. Hybrids present another situation. Whereas forms are the result of genetic diversity within a species, hybrids rely on an outside element, the pollinator, to create the resulting hybrid as a cross between two species. In certain parental combinations hybrids are very predictable, but in other situations they are very rare. Most are obvious intermediates between the putative parents, and when both parental species are present the hybrids should be carefully sought.

Flowering Periods

Flowering periods are not easy to isolate for this entire region. Typically plants flower earlier in the spring in northern Florida and later in the northern limit of our range and the upland mountain areas of the southern Appalachians, whereas autumn-flowering species tend to flower earlier in the north and later southward. A good example would be the **fragrant ladies'-tresses**, *Spiranthes odorata,* which flowers regularly in north Florida in October and November and in southern North Carolina in September. The seasons are noted and must be adjusted for the specific area in question. Flowering periods given for each species are intended to indicate the time of year that is most typical.

Range Maps

The range maps are intended to illustrate the general range of a given species. The maps are based on verifiable specimens in herbarium records, normally housed in herbaria at any number of colleges and universities. Upon rare occasion a verifiable report will be allowed. This would be a photograph from a reliable source, documented with date and place. Literature reports are noted in the text and are just that: a report in any one of a number of publications that cannot be backed up by either a specimen or verifiable record. The range maps do not attempt to differentiate between extant and extirpated populations, nor do they convey how many populations are known from a given area. If the range of a species or variety extends beyond the southeastern United States an arrow is used to indicate such.

Key to the Genera

1a plants epiphytic, or rarely lithophytic . . . *Epidendrum* p. 65
1b plants terrestrial . . . 2

2a lip inflated or sac shaped . . . 3
2b lip otherwise . . . 5

3a flowers small, whitish, and in spikes . . . *Goodyera,* p. 81
3b flowers otherwise . . . 4

4a lip a distinct slipper . . . *Cypripedium,* p. 53

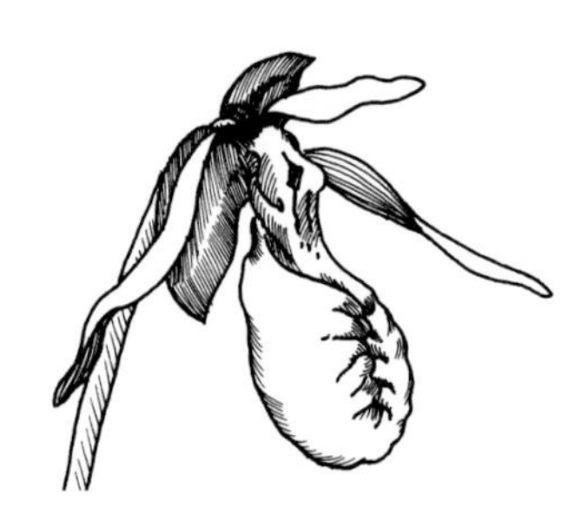

4b lip spade- or boat shaped . . . *Epipactis,* p. 69

Leaves Lacking at Flowering Time

5a (green) leaves (apparently) lacking at flowering time; stem bracts may be present . . . 6

5b (green) leaves present at flowering time . . . 12

6a plants lacking chlorophyll . . . 7

6b plants with chlorophyll . . . 8

7a flower with a mentum . . . *Corallorhiza,* p. 43

7b flower lacking a mentum . . . *Hexalectris,* p. 101

8a spur or mentum present . . . 9

8b spur or mentum lacking . . . 10

9a flowers asymmetrical . . . *Tipularia,* p. 231

9b flowers symmetrical . . . *Sacoila,* p. 187

10a flowers terminal, usually solitary (rarely 2–4) . . . *Arethusa,* p. 20

10b flowers numerous (usually 10+–60), arranged in a spike, often dense and/or spiraled . . . 11

11a inflorescence a spike, often with many (20–60) small, tubular, white or cream-colored flowers with delicately frilled lips, usually in a loose to dense spiral . . . *Spiranthes,* p. 190

11b inflorescence a raceme with fewer flowers (+10) . . . *Aplectrum,* p. 18

Leaves Present at Flowering Time

Spur or Mentum Present

12a spur or mentum present . . . 13

12b spur or mentum lacking . . . 20

13a inflorescence terminal on a leafy stem . . . 14

13b inflorescence terminal on a scape (leafless stem); flowers pink, purple, and/or white . . . *Galearis,* p. 77

14a lip oblong and notched at tip . . . *Coeloglossum,* p. 244

14b lip otherwise; *never* notched at the tip . . . 15

15a lip entire; the margin entire, erose, or toothed *and* lacking a prominent tubercle . . . *Gymnadeniopsis,* p. 85
15b lip otherwise . . . 16

16a plants otherwise; lip entire, divided, or fringed, but the margin *never* notched or toothed, *nor* ever forming colonies of basal rosettes . . . *Platanthera,* p. 134
16b lip otherwise . . . 17

17a petals deeply cleft or toothed, sterile plants forming distinctive colonies of large basal rosettes . . . *Habenaria,* p. 93
17b plants otherwise . . . 18

18a flowers white in short spikes, with a thick, bulbous spur . . . *Platythelys,* p. 171
18b flowers otherwise . . . 19

19a flowers green . . . *Platanthera,* p. 134
19b leaves basal with the scape arising from the center of the leaves . . . *Eulophia,* p. 73

Spur or Mentum Lacking

20a pseudobulbs present, although they may be well hidden in the leaf bases . . . 21

20b pseudobulbs lacking . . . 22

21a petals and sepals similar; inflorescence a slender spike or corymb . . . *Malaxis,* p. 123

21b petals and sepals dissimilar; petals filiform, threadlike . . . *Liparis,* p. 111

Leaves Basal

22a leaves essentially basal or extending up the lower 1/4 of the stem and rapidly reduced to leafy bracts . . . 23

22b leaves essentially cauline . . . 25

23a leaves numerous (5–10); flowers flat, held perpendicular to the axis on distinctive petioles . . . *Ponthieva,* p. 179

23b leaves few . . . 24

24a flowers white or cream in spikes, often spiraled . . . *Spiranthes,* p. 190
24b flowers copper-colored . . . *Mesadenus,* p. 131

Leaves Cauline

25a flowers non-resupinate . . . *Calopogon,* p. 25

25b flowers resupinate . . . 26

26a leaves opposite or whorled . . . 27
26b leaves solitary or alternate . . . 28

27a leaves 2, opposite . . . *Listera,* p. 117

27b leaves 5–7, whorled . . . *Isotria,* p. 105

28a leaves solitary, bracts may be present . . . 29
28b leaves multiple . . . 31

29a inflorescence a spike, raceme, or corymb of very small flowers . . . *Malaxis,* p. 123

29b inflorescence otherwise . . . 30

30a lip projecting forward, the petals inrolled to form a tube . . . *Cleistes,* p. 37

30b petals free, not inrolled; lip with a distinctive fringed beard . . . *Pogonia,* p. 175

31a inflorescence terminating a leafy stem; in a raceme, spike or cluster rarely individual . . . 32

31b inflorescence an axillary scape; flowers greenish yellow and purplish-black on tall scapes . . . *Pteroglossaspis,* p. 183

32a leaves plicate, lip striate. flowers purple . . . *Bletilla**, p. 22

32b leaves otherwise . . . 33

33a leaves small, oval, or scalelike . . . *Triphora,* p. 235

33b leaves linear, coppery green, flowers white with yellow on the lip . . . *Zeuxine**, p. 242

Part 2

Wild Orchids of the Southeastern United States

Aplectrum

Aplectrum is a small genus consisting of only two species, one found in North America, the other in Japan. It is one of several genera found in eastern North America that have analogs in Japan.

Aplectrum hyemale (Mühlenberg *ex* Willdenow) Nuttall

putty-root, Adam-and-Eve

Minnesota east to Massachusetts, south to northern South Carolina, and west to eastern Oklahoma

forma *pallidum* House—yellow-flowered form

Torreya 3:54. 1903 as *Aplectrum spicatum* var. *pallidum* House, type: New York

Alabama, Georgia, South Carolina: rare to occasional, more frequent northward; found primarily in the mountains and Piedmont
Plant: terrestrial, 18–50 cm tall
Leaves: 1; ovate, plicate, 3–8 cm wide × 10–20 cm long; appearing in the autumn and withering at flowering time in the spring
Flowers: 3–20, greenish-yellow suffused with brown and purple, the white lip 3-lobed, prominently ridged, and with magenta spots or, in the forma *pallidum*, the stem and flowers yellow and the lip unspotted; individual flower size 1.5–2.0 cm
Habitat: rich, moist woodlands, bottomlands, and deciduous slopes
Flowering period: April–mid-May

Aplectrum, a not uncommon orchid of the central and southern Appalachian Mountains and foothills, reaches the southern limit of its range in northern Alabama, Georgia, and South Carolina. It is nowhere frequent in the Southeast and, although the flowers are prominent, the coloring renders them as almost inconspicuous in the similarly colored forest surroundings. The distinctive leaf is much easier to find in the late autumn of the year when it is nearly erect. By spring it has reclined and is usually withering when the flower spike starts to emerge. The large, pendant seed capsules also are diagnostic on the fruiting stems later in the season.

Arethusa

Arethusa is a monotypic genus found in eastern North America and Japan. The brilliantly colored flowers are a feature of many of the bogs and fens of northeastern North America. The genus reaches the extreme limit of its range in northern North Carolina.

Arethusa bulbosa Linnaeus

dragon's-mouth

southern Manitoba east to Newfoundland, and south to northern South Carolina, west to northern Indiana, and central Minnesota

forma *albiflora* Rand & Redfield—white-flowered form

Flora of Mt. Desert Island 152. 1894, type: Maine

forma *subcaerulea* Rand & Redfield—lilac-blue-flowered form

Flora of Mt. Desert Island 152. 1894, type: Maine

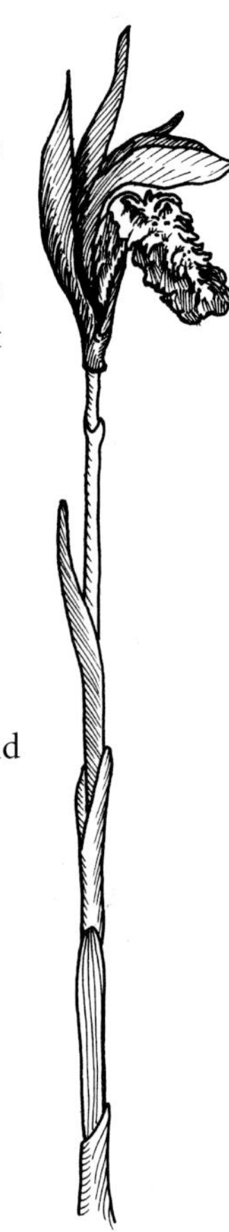

South Carolina: very rare in the northwestern bogs and seeps of Pickens and Greenville Counties; no confirmed sightings for many years
Plant: terrestrial, 4–15(20) cm tall
Leaves: 1; linear, 3–12 mm wide and 5–23 cm long; appressed to the flowering stem when young and continuing to develop as the plant matures
Flowers: usually 1, occasionally 2–4; rose, rich pink, or magenta or, in the forma *albiflora*, flowers pure white or, in the rarer forma *subcaerulea*, lilac-blue; individual flower size 1–3 cm; miniature individuals, with flowers no more than 0.5 cm tall, rarely occur
Habitat: sphagnum bogs, fens, and seeps
Flowering period: late spring

To many, *Arethusa* is the ultimate gem of the northern affinity bogs and fens. It is one of the rarest orchids in the Southeast and is only known from a few isolated sites in northern South Carolina where it may not have been seen in many years. Plants often appear leafless at flowering time, as the emerging grasslike leaf is nearly appressed to the flower stalk, although it will elongate later in the season. The beautiful pink flowers are distinctive and could only be confused (at a distance) with the **rose pogonia**, *Pogonia ophioglossoides*, with which it frequently grows.

forma *albiflora*

Bletilla

Bletilla is a genus of hardy Asian orchids that has become increasingly popular in North American gardens. Plants are easily obtained in local garden centers and via mail order catalogues. Because they propagate easily by rhizome pieces, plants are often traded with other gardeners. A single population has been found in the wild in Florida

Bletilla striata (Thunberg) Reichenbach *f.**

urn orchid

Asia
Florida: a single record from Escambia County; native to China
Plant: terrestrial, 10–50 cm tall
Leaves: 3–5; ovate, plicate, strongly ribbed, 1.5–3.0 cm wide and to 20 cm long
Flowers: 5–12, most of which are open simultaneously; rich magenta/pink; individual flower size 2.0–2.5 cm with a prominent ruffled crest
Habitat: wooded roadside as a garden escape
Flowering period: late February–April
Bletilla is a popular garden plant that has appeared in the wild and is persisting. Its occurrence may have resulted from rhizome pieces discarded in garden waste.

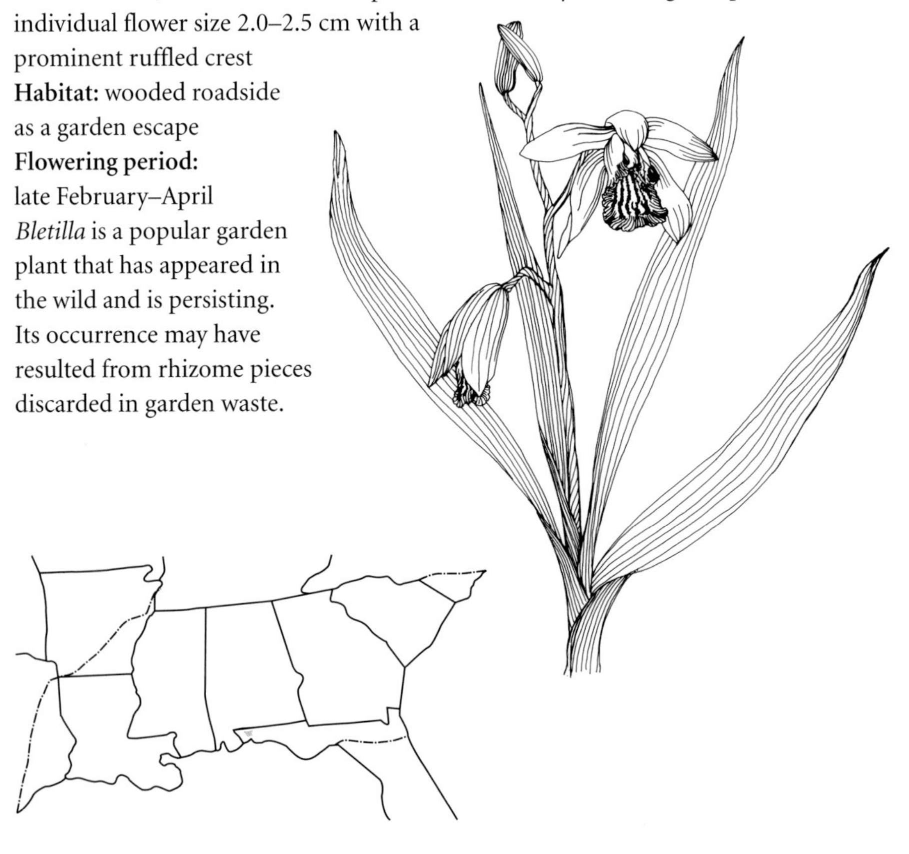

Calopogon

The genus *Calopogon* is a New World genus composed of five species, only one of which also occurs outside of the United States and Canada. All five of these species are found in the southeastern United States. The non-resupinate (uppermost) lip is distinctive and easily identifies the genus.

1a petals widest above the middle . . . **many-flowered grass-pink**, *C. multiflorus*
1b petals equal to or widest below the middle . . . 2

2a leaf appressed to inflorescence, flowers approximate, opening simultaneously . . . **bearded grass-pink**, *C. barbatus*
2b leaf not appressed to inflorescence, flowers more widely spaced . . . 3

3a flowers opening nearly simultaneously, floral bracts 4–8 mm, plants of prairie habitats . . . **Oklahoma grass-pink**, *C. oklahomensis*
3b flowers opening sequentially over a period of time . . . 4

4a petals narrow, falcate, and strongly ascending . . . **pale grass-pink**, *C. pallidus*
4b petals broad and spreading . . . **common grass-pink**, *C. tuberosus*

Calopogon barbatus (Walter) Ames

bearded grass-pink

North Carolina south to Florida and west to Louisiana
  forma *albiflorus* P. M. Brown—white-flowered form
    North American Native Orchid Journal 9: 33. 2003, type: North Carolina
  forma *lilacinus* P. M. Brown—lilac-flowered form
    North American Native Orchid Journal 9: 33. 2003, type: Georgia

Louisiana, Alabama, Mississippi, Georgia, Florida, South Carolina, North Carolina: locally abundant in the southern portion of its range and becoming scattered and rare on the periphery; extends south in Florida beyond our limits
Plant: terrestrial, 5–20 cm tall
Leaves: 1 or 2; slender, 0.5–1.0 cm wide and to 20 cm long
Flowers: 3–7, non-resupinate, most of which are open simultaneously; bright magenta-pink or, in the forma *albiflorus,* pure white or, in the forma *lilacinus,* bluish-lilac; individual flower size 2.0–2.5 cm
Habitat: wet meadows, pine flatwoods, and sphagnous roadsides
Flowering period: late March–May

This is the earliest of the grass-pinks to flower and is similar to the very rare **many-flowered grass-pink**, *Calopogon multiflorus,* but grows in a much wider variety of habitats and can be separated from the latter by the shape of the petals. The petals are wider below the middle on *C. barbatus;* it usually flowers earlier than the **common grass-pink**, *C. tuberosus,* and is a much smaller plant. In Louisiana plants should be examined carefully for the possibility of *C. oklahomensis.* The bearded grass-pink, *C. barbatus,* is often found in the company of a variety of carnivorous plants—pitcherplants, *Sarracenia* spp., sundews, *Drosera* spp., and butterworts, *Pinguicula* spp.

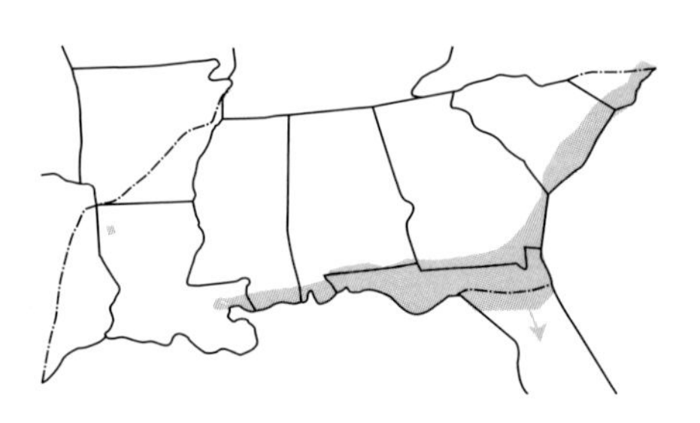

forma *lilacinus*

forma *albiflorus*

Calopogon multiflorus Lindley

many-flowered grass-pink

North Carolina south to Florida and west to eastern Louisiana
Louisiana, Alabama, Mississippi, Georgia, Florida, South Carolina, North Carolina: rare and local throughout; extending south in Florida
Globally Threatened
Plant: terrestrial, 15–30 cm tall
Leaves: 1 or 2; slender, 3 mm wide × 10 cm long and less than the height of the plant
Flowers: 5–15, non-resupinate, most of which are open simultaneously; bright magenta-pink with a golden crest on the lip;
individual flower size 2.0–2.5 cm
Habitat: damp meadows, pine flatwoods
Flowering period: (March) April–May (July)

Calopogon multiflorus is the rarest of the grass-pinks to be found in the United States and has declined dramatically over the last 25 years. It is primarily a fire-respondent species and often does not flower until a few weeks following a spring burn. Most populations in our range consist of only a few plants. Hybrids with *C. pallidus* have been reported from Florida by Goldman (2000, 2002).

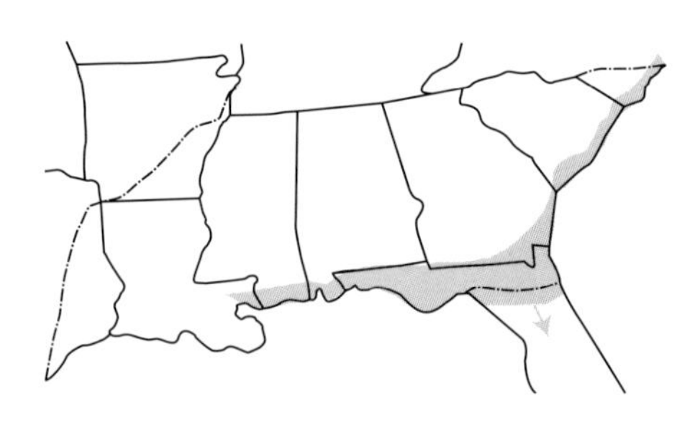

Calopogon oklahomensis D. H. Goldman

Oklahoma grass-pink

southern Minnesota east to western Indiana, south to southern Georgia, and west to eastern Texas

forma *albiflorus* P. M. Brown—white-flowered form

North American Native Orchid Journal 9: 33–34. 2003, type: Arkansas

Texas, Louisiana, Arkansas, Alabama, Mississippi, Georgia, South Carolina: local in remnant prairies; historical elsewhere
Plant: terrestrial, 15–36 cm tall
Leaves: 1 or 2; lanceolate, slender, 0.5–1.5 cm wide × 7–35 cm long
Flowers: 3–7(13), non-resupinate, most of which are open simultaneously; color is highly variable from lilac-blue to bright magenta-pink with a golden crest on the lip or, in the forma *albiflorus*, white; individual flower size 2.5–4.0 cm
Habitat: prairies, pine savannas, open flatwoods, and frequently mowed damp meadows
Flowering period: (March) April–May (June)

Calopogon oklahomensis is the most recent of the grass-pinks to be described from the United States. Originally thought to be restricted to the prairies of the south-central states, it is now known to be considerably more widespread although extirpated from much of the original range. Spring in the prairies of eastern Arkansas often brings a tapestry of color, with hundreds of the Oklahoma grass-pinks in various shades of pink. The South Carolina record is based upon a specimen without precise location data. Local prairielike habitats exist in the sandhills of the Piedmont in South Carolina, often with other species typical of the Midwestern prairies. Plants previously identified as *Calopogon barbatus* or small, early flowering *C. tuberosus* occurring in suitable habitat should be carefully examined for the possibility of *C. oklahomensis*. Goldman (2000) suggests that this species may have arisen from ancient hybridization of *C. barbatus* and *C. tuberosus*.

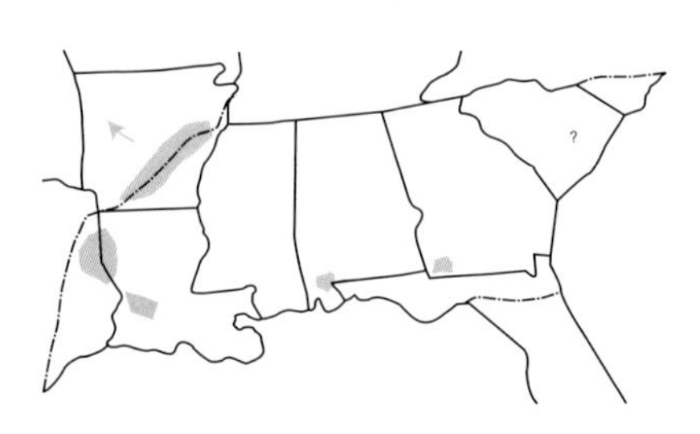

forma *albiflorus*

Calopogon pallidus Chapman

pale grass-pink

southeastern Virginia south to Florida and west to Louisiana
forma *albiflorus* P. M. Brown—white-flowered form
North American Native Orchid Journal 1(1): 8. 1995, type: Florida

Alabama, Mississippi, Georgia, Florida, South Carolina, North Carolina: occasional to frequent in much of its range; confined to Coastal Plain areas and southward in Florida beyond our area
Plant: terrestrial, 15–50 cm tall
Leaves: 1 or 2; slender, ribbed, 0.5 cm wide × 10–20 cm long and less than the height of the plant
Flowers: 5–12, non-resupinate, opening in slow succession; color can be highly variable but typically pale magenta-pink with a golden crest on the lip or, in the forma *albiflorus*, white; individual flower size 2.0–3.5 cm
Habitat: wet meadows, pine flatwoods, and sphagnous roadsides
Flowering period: March–July

Second in abundance to the common grass-pink, *Calopogon pallidus* is widespread through most of the Coastal Plain of the southeastern United States. It often grows in habitats with other species of grass-pinks and the distinctive upwardly curved petals make it easy to separate from those other species. Despite its specific epithet, the flower color is not always pale and can vary from white to a rich, deep magenta. Hybrids with *C. multiflorus* have been reported by Goldman (2000, 2002).

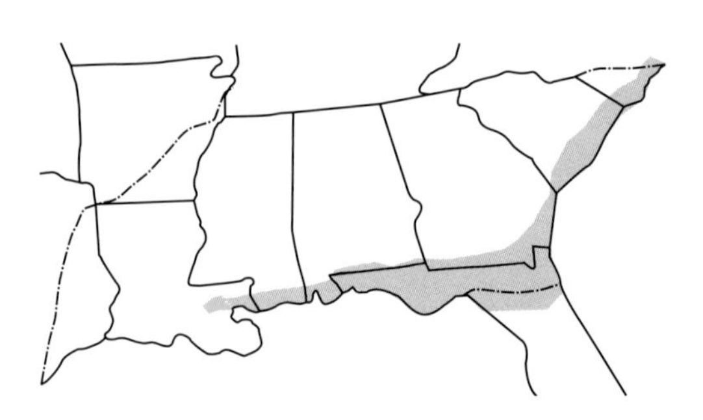

forma *albiflorus*

Calopogon tuberosus (Linnaeus) Britton, Sterns, & Poggenberg var. *tuberosus*

common grass-pink

Minnesota east to Newfoundland, south to Florida, and west to Texas
forma *albiflorus* Britton—white-flowered form
Bulletin of the Torrey Botanical Club 17: 125. 1890, type: New Jersey

Texas, Louisiana, Arkansas, Alabama, Mississippi, Georgia, Florida, South Carolina, North Carolina: occasional to frequent in all states
Plant: terrestrial, 25–115 cm tall
Leaves: 1–2(3); slender, ribbed, up to 0.3–4.0 cm wide × 3–45 cm long and less than the height of the plant
Flowers: 3–17, non-resupinate, opening in slow succession; deep to pale pink with a golden crest on the lip or, in the forma *albiflorus*, white; individual flower size 2.5–3.5 cm
Habitat: wet meadows, pine flatwoods, open prairies, mountain bogs, seeps, and sphagnous roadsides
Flowering period: March–July

One of the most frequent orchids to be found in the eastern and central United States and Canada, the brilliant, showy *Calopogon tuberosus* prefers open, wet, sandy roadsides, sphagnum bogs, and seeps. Plants of the common grass-pink flower over a very long period of time, with only a few flowers open at once. It is not unusual to find local stands of several hundred plants. Like all grass-pinks, the flowers have the lip uppermost, non-resupinate, and this feature easily separates the genus from any other with a similar morphology. The var. *simpsonii*, found in rocky marls of the southernmost counties of Florida, does not occur within our range of the Southeast.

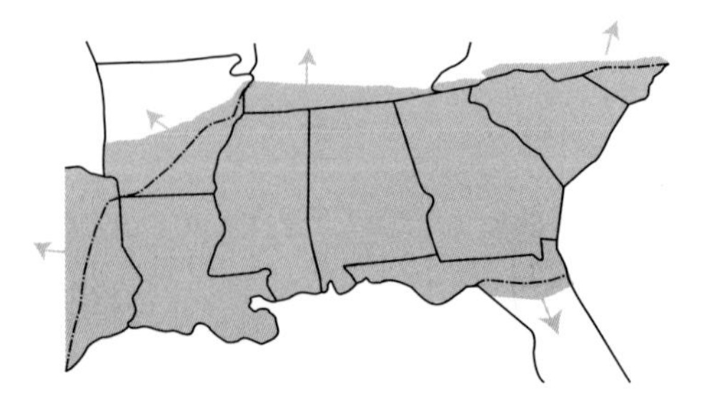

forma *albiflorus*

Cleistes

The genus *Cleistes* is composed of about 25 species in the Western Hemisphere, many of which grow in wet savannas in northern South America. Most of the species have, at one time, been classified as *Pogonia*, a very closely related genus, and some current authors treat them that way today. Based on recent DNA analyses, the two species in the United States may properly belong in a new genus.

1a column 13–19 mm long; lip 21–33 mm long; leaf and bract broadly lanceolate . . . **upland spreading pogonia**, *C. bifaria*

1b column 21–25 mm long; lip 34–56 mm long; leaf and bract narrowly lanceolate . . . **spreading pogonia**, *C. divaricata*

Cleistes bifaria (Fernald) Catling & Gregg

upland spreading pogonia

West Virginia south to Florida, west to eastern Louisiana and eastern Texas
Texas, Louisiana, Alabama, Mississippi, Georgia, Florida, South Carolina, North Carolina: rare and local throughout its range and extending to the uplands of Tennessee, North Carolina, and West Virginia; extirpated in Texas
Plant: terrestrial, 20–50 cm tall
Leaves: 1; broadly lanceolate, glaucous, 3 cm wide × 15 cm long; a smaller floral bract, 1.5 × 6.5 cm, resembling a leaf, subtends the flower
Flowers: 1, rarely 2; sepals bronzy-green, petals pale pink to blush; lip whitish veined in darker pink with a yellow crest; lip to 21–33 mm long and appearing with a broadened apex; individual flower size 5 × 5 cm
Habitat: open pine flatwoods, shaded prairies, and seepy meadows
Flowering period: April–May

Both species of *Cleistes* are two of our showiest orchids in the southeastern United States. Although range overlap is minimal they can both be found in the same site in a few locations in North Carolina and Florida. Size (bract and lip width), coloration, and flowering time are all helpful criteria in determining identification. *Cleistes bifaria* appears to prefer a wetter habitat than *C. divaricata*, flowers earlier, and is generally paler in color. The only absolute criteria are those in the key and require measurements. After a few seasons of experience it becomes quite easy to tell them apart at a glance. *Cleistes bifaria* does not appear to be as fire-dependent as *C. divaricata.*

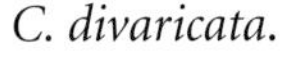

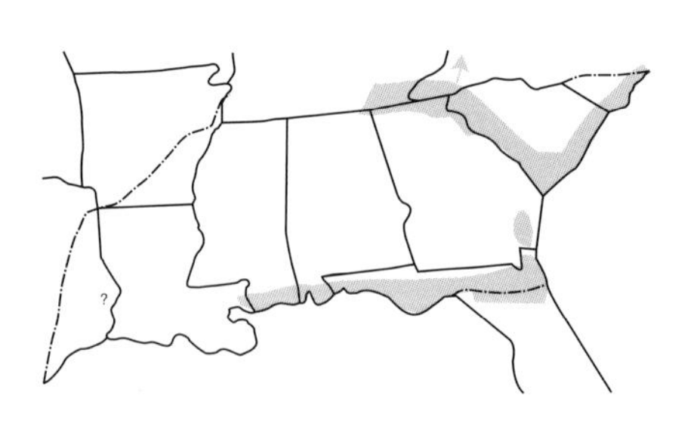

Cleistes divaricata (Linnaeus) Ames

spreading pogonia

New Jersey south to Florida; primarily along the Coastal Plain
forma *leucantha* P. M. Brown—white-flowered form
North American Native Orchid Journal 1(1): 8. 1995, type: Florida

Georgia, Florida, South Carolina, North Carolina: occasional to locally abundant along the Coastal Plain; historical disjunct locations extend northward to central New Jersey
Plant: terrestrial, 30–70 cm tall
Leaves: 1; narrowly lanceolate, glaucous, to 2.5 cm wide × 20.0 cm long; a smaller floral bract, 1.5 × 8.0 cm, resembling a leaf, subtends the flower
Flowers: 1, rarely 2; sepals purple, petals deep pink; lip pink veined in darker pink with a yellow crest or, in the forma *leucantha*, the sepals apple green and the petals white; lip to 34–56 mm long, appearing to taper to a point; individual flower size 7 × 7 cm
Habitat: open pine flatwoods, shaded prairies, and damp meadows; usually in recently burned areas
Flowering period: April–May

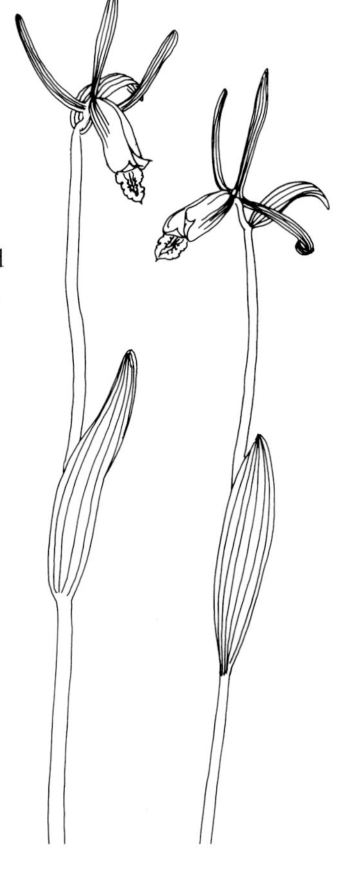

The large spreading pogonia is unquestionably one of the most highly sought after native orchids found on the southeastern Coastal Plain. The large, showy, rosy-pink flowers hold their heads high among the vegetation. Plants seem to flower best two to three years following a burn. Subsequently the competition becomes too dense and only sterile leaves can be found.

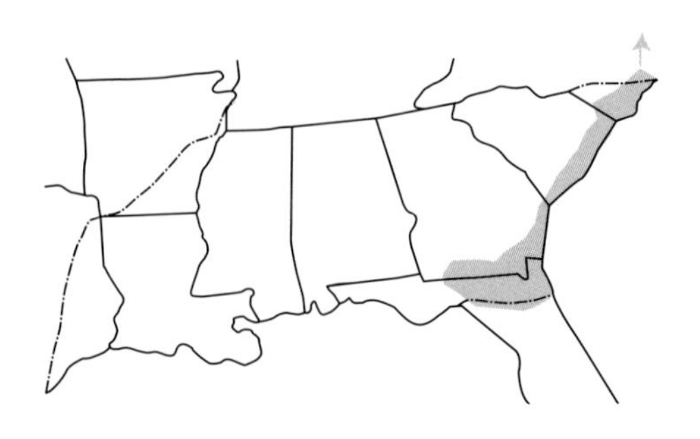

forma *leucantha*

Corallorhiza

The genus *Corallorhiza* has 13 species throughout North America and Hispaniola. One species, *Corallorhiza trifida,* is widespread across Eurasia. The plants are entirely mycotrophic and some are thought to be saprophytes. They arise from a coralloid rhizome, hence the name. The entire genus is easily recognizable from its leafless stems, although they may be variously colored, and by their small flowers. Three species and one variety are found within the southeastern United States.

1a flowers cleistogamous, very small, less than 3 mm; autumn flowering . . . **autumn coralroot**, *C. odontorhiza* var. *odontorhiza*
1b flowers chasmogamous, the lip clearly defined . . . 2

2a winter and spring flowering . . . **Wister's coralroot**, *C. wisteriana*
2b summer and autumn flowering . . . 3

3a petals and sepals distinct, late spring-summer flowering . . . **spotted coralroot**, *C. maculata*
3b petals and sepals indistinct, autumn flowering . . . 4

4a lip prominent, flowers chasmogamous . . . **Pringle's autumn coralroot**, *C. odontorhiza* var. *pringlei*
4b lip not prominent, often undeveloped . . . **autumn coralroot**, *C. odontorhiza* var. *odontorhiza*

Corallorhiza maculata (Rafinesque) Rafinesque var. *maculata*

spotted coralroot

British Columbia east to Newfoundland, south to California, Arizona, and New Mexico; in the Appalachian Mountains south to northern Georgia and South Carolina

forma *flavida* (Peck) Farwell—yellow-stemmed form

Report (Annual) of the Regents University of the State of New York. New York State Museum 50: 126. 1897 as *Corallorhiza multiflora* var. *flavida* Peck, type: New York

forma *rubra* P. M. Brown—red-stemmed form

North American Native Orchid Journal 1(1): 8–9. 1995, type: Vermont

Georgia, South Carolina: very rare and local at the southern limits of its range; range extends both west and northward throughout much of North America
Plant: terrestrial, mycotrophic, 20–50 cm tall; stems bronzy-tan or, in the forma *flavida*, bright yellow or, in the forma *rubra*, red
Leaves: none
Flowers: 5–20; tepals typically brownish or, in the forma *flavida*, bright yellow or, in the forma *rubra*, red; lip white, spotted with madder purple; in the forma *flavida*, unspotted or, in the forma *rubra*, spotted with bright red; individual flowers 5.0–7.5 mm
Habitat: rich mesic and mixed forests
Flowering period: late May–July

Although the spotted coralroot is the most frequently encountered species of coralroot found within eastern North America, it is reaching the extreme southern limit of its range in northern Georgia and South Carolina. Elsewhere the variation in the stem color is evident but in the Southeast, because so few plants are to be found, such variation may not be as apparent. *Corallorhiza maculata* var. *occidentalis*, the western spotted coralroot, occurs primarily in the western and northern portions of North America and *C. maculata* var. *mexicana* in southwestern Arizona and Mexico.

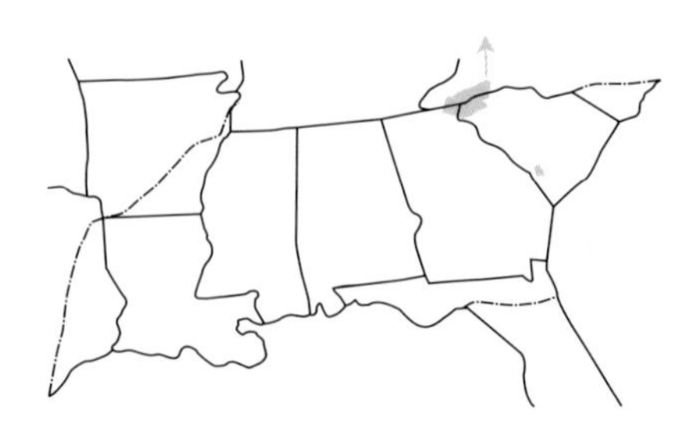

forma *rubra*

forma *flavida*

Corallorhiza odontorhiza (Willdenow) Poiret var. *odontorhiza*

autumn coralroot

South Dakota east to Maine, south to Oklahoma and northern Florida
forma *flavida* Wherry—yellow-stemmed form
Journal of the Washington Academy of Science 17: 36. 1927, type: Washington, D.C.

Texas, Louisiana, Arkansas, Alabama, Mississippi, Georgia, Florida, South Carolina, North Carolina: widespread, but never common—this plant is both rare and easily overlooked
Plant: terrestrial, mycotrophic, 5–10 cm tall; stems bronzy-green or, in the forma *flavida*, yellow
Leaves: lacking
Flowers: 5–12; cleistogamous; sepals green suffused with purple, covering the petals; lip, rarely evident in this variety, white spotted with purple or, in the forma *flavida*, unspotted; individual flower size 3–4 mm
Habitat: rich, calcareous woodlands
Flowering period: September–October

The fact that this inconspicuous little orchid is rarely found may be attributed more to its size and habit than necessarily to its rarity. The autumn coralroot appears to be never common anywhere and is usually found by accident. The short stems often flower among the fallen leaves in the autumn months and the coloration, *sans* chlorophyll, makes them even harder to see. In Florida it is known from only a single location, where it reaches the southern limit of its range in Columbia County.

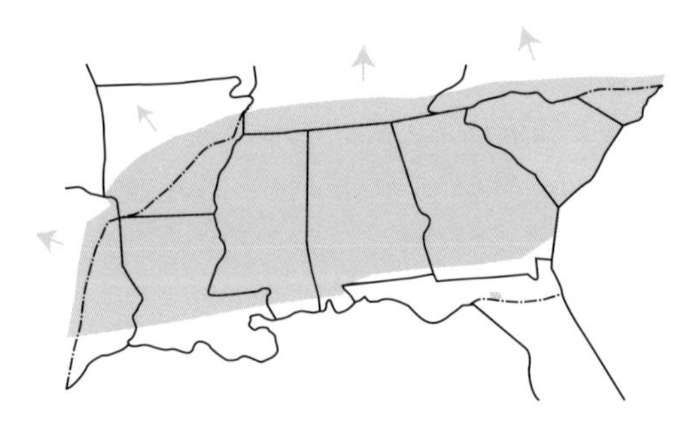

Corallorhiza odontorhiza (Willdenow) Poiret var. *pringlei* (Greenman) Freudenstein

Pringle's autumn coralroot

Wisconsin and Ontario east to Maine, south to Iowa, Tennessee, and Georgia; Mexico, Central America
Georgia: recorded from a single site in the south-central part of the state
Plant: terrestrial, mycotrophic, 5–15 cm tall; stems bronzy-green
Leaves: lacking
Flowers: 5–12; chasmogamous; the sepals green suffused with purple, spreading and revealing the petals; lip broad, white, spotted with purple; individual flower size 5–10 mm
Habitat: rich, mesic forests and calcareous woodlands
Flowering period: September–October

Pringle's autumn coralroot is an extremely rare variety with its center of distribution in the lower Great Lakes states. This variety has only been recently described, although plants with showy, chasmogamous flowers have been known for some time. In the Southeast it is known from a single collection from northeast Georgia.

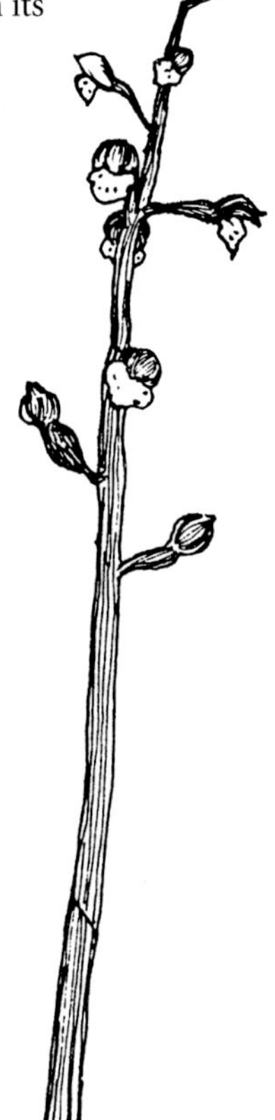

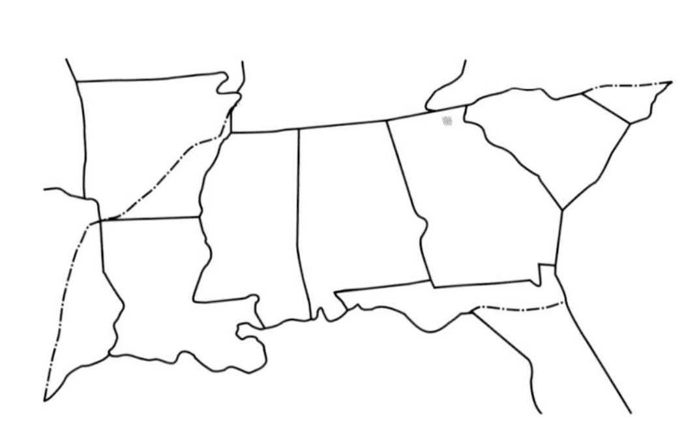

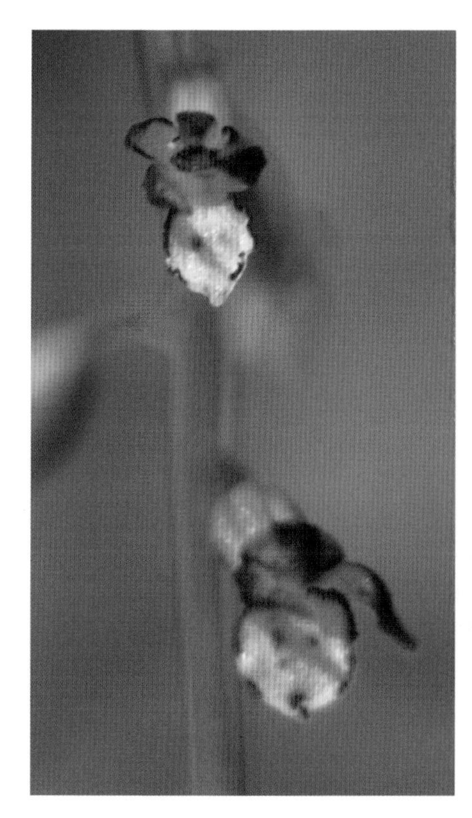

Corallorhiza wisteriana Conrad

Wister's coralroot

Washington east to New Jersey, south to Arizona and Florida; Mexico
forma *albolabia* P. M. Brown—white-lipped form
North American Native Orchid Journal 1(1): 9–10. 1995, type: Florida
forma *rubra* P. M. Brown—red-stemmed form
North American Native Orchid Journal 1(1): 62. 2000, type: Florida

Texas, Louisiana, Arkansas, Alabama, Mississippi, Georgia, Florida, South Carolina, North Carolina: widespread and locally abundant; range extends both north and west of our area and southward in Florida
Plant: terrestrial, mycotrophic, 5–30 cm tall; stem brownish-yellow or, in the forma *albolabia*, yellow or, in the forma *rubra,* red
Leaves: lacking
Flowers: 5–25; sepals green, petals yellow suffused and mottled with purple; lip white spotted with purple or, in the forma *albolabia*, yellow-stemmed, sepals and petals yellow with a pure white lip or, in the forma *rubra*, red-stemmed, sepals and petals red, with flowers marked red; individual flower size 5–7 mm
Habitat: rich, often calcareous woodlands, pine flatwoods, occasionally in lawns and foundation plantings
Flowering period: late December–April

The first truly native orchid to flower in the new year in the southeastern United States is often Wister's coralroot. The pale brown stems and small, spotted, white flowers can usually be seen in early January in Florida and on into March or early April northward. It occurs in scattered locations in open woods and even homesites. Although it can be found as an individual, some sites have several thousand plants in large, clustered colonies. The two color forms are exceedingly rare and known from very few sites.

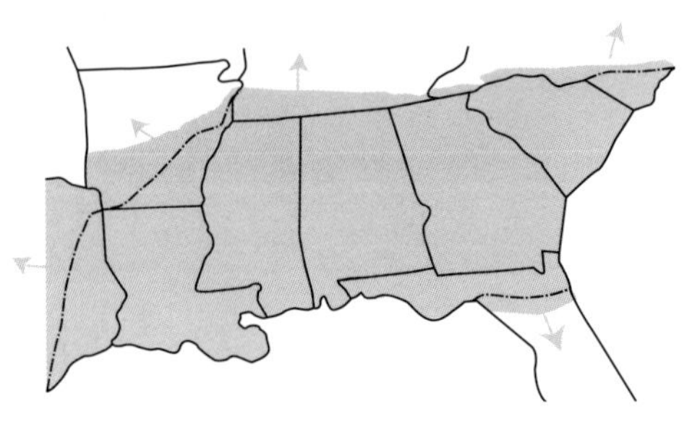

forma *rubra*

forma *albolabia*

Cypripedium

Cypripedium is a distinctive genus of about 45 species with 12 occurring in North America, north of Mexico. Although the leaf arrangement is variable, the lip, an unmistakable pouch-shaped slipper, is always diagnostic. This is often the genus that is first recognized by orchid enthusiasts. There are three species found within the southeastern United States and one, the showy-lady's-slipper, is found just north of our area in North Carolina.

1a leaves basal . . . **pink lady's-slipper**, *C. acaule*
1b leaves cauline . . . 2

2a lip orbicular, flowers very large; white, ivory, or yellow . . . **ivory-lipped lady's-slipper**, *C. kentuckiense*
2b lip oval to oblong . . . 3

3a petals marked with purple, lip white, streaked with lavender . . . **small white lady's-slipper**, *C. candidum*
3b lip yellow . . . 4

4a flowers commonly large, lip to 5.4 cm long; sepals and petals unmarked to spotted, striped, or reticulately marked with reddish-brown or madder; plants of a variety of habitats, usually mesic to calcareous woodlands or open sites in limestone or gypsum . . . **large yellow lady's-slipper**, *C. parviflorum* var. *pubescens*
4b flowers small, lip 2.2–3.4 cm long; sepals and petals usually densely spotted with dark reddish-brown appearing as a uniformly dark wash; plants of dry, deciduous, more acidic sites than *C. parviflorum* var. *pubescens* . . . **southern small yellow lady's-slipper**, *C. parviflorum* var. *parviflorum*

Cypripedium acaule Aiton

pink lady's-slipper, moccasin flower

Northwest Territories east to Newfoundland, south to Minnesota, Mississippi, and Georgia

forma *albiflorum* Rand & Redfield—white-flowered form

Flora of Mt. Desert Island 154. 1894, type: Maine

forma *biflorum* P. M. Brown—2-flowered form

North American Native Orchid Journal 1: 197. 1995, type: New Hampshire

Alabama, Mississippi, Georgia, South Carolina, North Carolina: locally frequent in the mountains and Piedmont areas east of the Mississippi River
Plant: terrestrial, 10–55 cm tall
Leaves: 2; 5–13 cm wide × 10–30 cm long; pubescent
Flowers: 1, rarely 2 in the forma *biflorum*; sepals green–reddish-brown, petals bronze; lip pale rosy-pink–deep raspberry or, in the forma *albiflorum*, white with pale green petals and sepals; individual flower size ca. 4 × 4 cm; lip 3–6 cm long with a longitudinal fissure
Habitat: mixed hardwood and coniferous forest; usually in highly acidic soils
Flowering period: spring, April–May

Perhaps one of the most familiar orchids to be found in eastern North America, *Cypripedium acaule* reaches the southern limit of its range in the northern counties of the Southeast. Although color is variable and presents itself in just about every shade of pink, some actually tend toward peach. The forma *albiflorum* is rare and has been seen in Alabama. Plants are notoriously difficult to transplant and reestablish, although they may persist for a few years. Resist the temptation to move plants, unless they are to be destroyed. It is far better to just admire them in their natural surroundings.

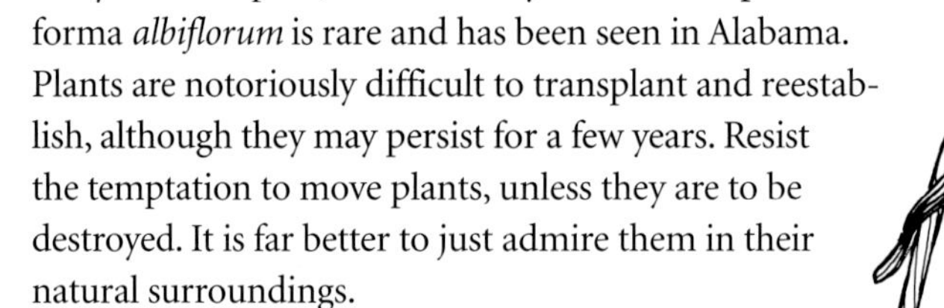

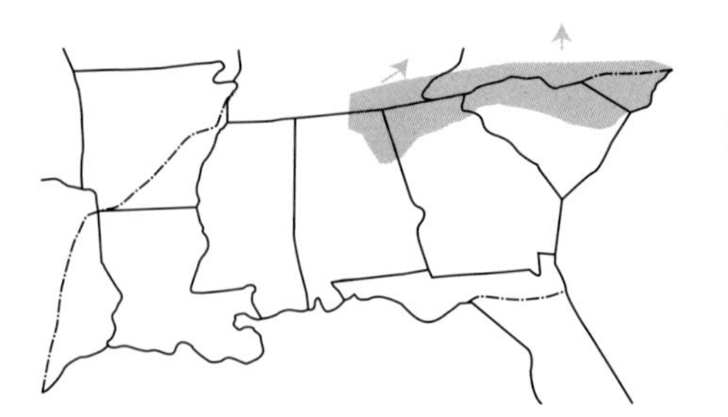

forma *albiflorum*

Cypripedium candidum Mühlenberg *ex* Willdenow

small white lady's-slipper

Saskatchewan south to Nebraska, east to western New York, south to Missouri, Kentucky, and New Jersey; Alabama
Alabama: very rare; two extralimital populations have been recently found in northern and central Alabama
Globally Threatened
Plant: terrestrial, 11–30 cm tall
Leaves: 3–5; alternate, ascending, elliptic to oblanceolate, 2–5 cm wide × 10–20 cm long
Flowers: 1–3; sepals and petals green to ochre striped with magenta, lateral sepals united; petals undulate and spiraled; lip white with delicate lavender striping beneath; individual flower size ca. 4 × 3 cm; lip 1.5–2.6 cm, the opening ovate at the base of the lip
Habitat: calcareous fens and prairies
Flowering period: early spring; March in our area

The small white lady's-slipper, one of the most imperiled of all the lady's-slippers in North America, was recently found in two small populations in Alabama. This is a considerable distance from the nearest populations in southwestern Kentucky. Precise locations have not been disclosed to the general public for fear of theft. When in flower this is the shortest in stature of the eastern lady's-slippers and one of the most highly scented. A delicate, but distinct, sweet odor can be detected from as far away as a meter. When the plants emerge the flower buds are tightly enclosed within the unfolding leaves and then pop out to open while the leaves are still usually clasping. After flowering the leaves then fully expand, although they still remain erect.

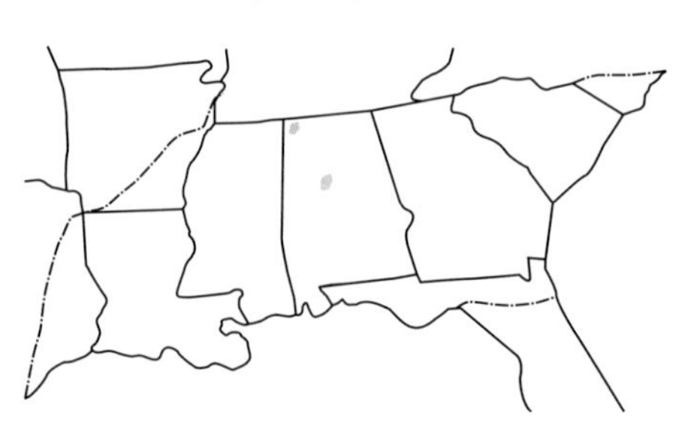

Cypripedium kentuckiense C. F. Reed

ivory-lipped lady's-slipper

northeastern Virginia; Kentucky south to eastern Texas and east to Georgia

forma *pricei* P. M. Brown—white-flowered form

North American Native Orchid Journal 4: 45. 1998, type: Arkansas

forma *summersii* P. M. Brown—concolorous yellow-flowered form

North American Native Orchid Journal 8: 30–31. 2002, type: Arkansas

Texas, Louisiana, Arkansas, Alabama, Mississippi, Georgia: rare and local in most areas with a nearly disjunct population in east-central Georgia; additional populations exist to the west in Oklahoma, north to Tennessee and Kentucky and a disjunct population in northeastern Virginia

Plant: terrestrial, 35–98 cm tall

Leaves: 3–6; alternate, evenly spaced along the stem, broadly ovate to ovate-elliptic, 5–13 cm wide × 10–24 cm long

Flowers: 1–2(3); sepals and petals green to yellowish-green prominently striped or spotted with dark reddish-brown or, in the forma *pricei*, pale green or, in the forma *summersii*, yellow, the lateral sepals united; petals undulate and spiraled to 15 cm long; lip ivory or pale to deep yellow with delicate green spotting within or, in the forma *pricei*, white or, in the forma *summersii*, yellow and unspotted; individual flower size ca. 20 × 15 cm; lip 5.0–6.5 cm, the opening ovate at the base of the lip, with the overall lip compressed laterally unlike any other of our lady's-slippers

Habitat: deciduous wooded seeps, alluvial forests, bases of slopes; occasionally in adjacent floodplains

Flowering period: late spring; mid-April–early May in our area

Cypripedium kentuckiense is the largest flowered of all of the lady's-slippers to be found in the United States and certainly one of the showiest. The enormous flowers are nearly twice the overall size of any of the other yellow-flowered lady's-slippers. Colonies in the eastern portion of our range are small and scattered whereas in Louisiana, Arkansas, and eastern Texas the colonies are much larger and thriving. The recent discovery of a small colony in east-central Georgia is the southeastern limit of the range for this spectacular species.

forma *pricei*

forma *summersii*

Cypripedium parviflorum Salisbury var. *parviflorum*

southern small yellow lady's-slipper

Kansas east to Massachusetts, south to Arkansas and Georgia

forma *albolabium* Magrath & Norman—white-lipped form

Sida 13: 372. 1989, type: Oklahoma

Arkansas, Alabama, Georgia: rare and local in the northern and central deciduous forests; extensive habitat exists in Mississippi and South Carolina
Plant: terrestrial, 10–60 cm tall
Leaves: 4–5; alternate, evenly spaced along the stem, spreading; ovate to ovate-elliptic to lance-elliptic, 2.5–8.0 cm wide × 10–18 cm long
Flowers: 1–2(3); sepals and petals uniformly dark with dense, minute dark chestnut or reddish-brown spots, often appearing as a uniform color; lateral sepals united; petals undulate and spiraled to 10 cm long; lip slipper-shaped, deep, rich yellow, often with scarlet markings within the lip or, in the forma *albolabium*, white; individual flower size ca. 4.5 × 10.0 cm; lip 2.2–3.4 cm, the opening ovate-oblong at the base of the lip
Habitat: decidedly acid, deciduous, dry to mesic wooded slopes
Flowering period: late spring; mid-April–early May in our area

The recent separation of the two small yellow lady's-slippers has solved an often-confusing question. How can small-flowered, dark-petalled yellow lady's-slippers that occur in dry, acid woods be the same as the similar plants in the typical calcareous woodlands and wetlands? The answer is that they are not the same; nor are they gradations of the large yellow. The habitat, range, and delicate roselike fragrance should be more than sufficient to identify this uncommon variety. Although vouchers apparently do not exist for Mississippi and South Carolina, plants of *Cypripedium parviflorum* var. *pubescens* occurring in acid woodlands should be carefully examined for the possibility of *C. parviflorum* var. *parviflorum*.

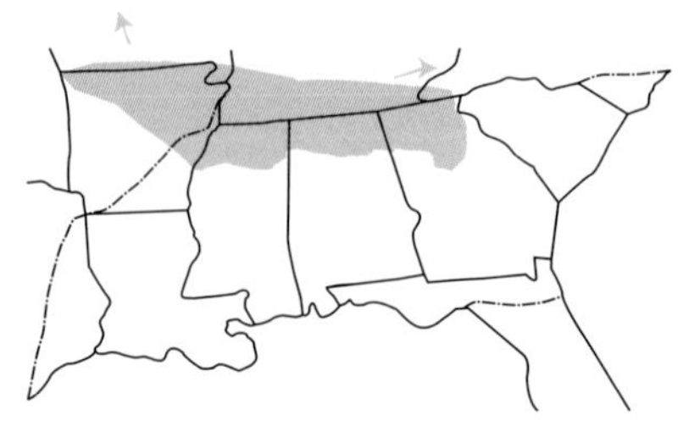

Cypripedium parviflorum Salisbury var. *pubescens* (Willdenow) Knight

large yellow lady's-slipper

Alaska east to Newfoundland, south to Arizona and Georgia
Alabama, Georgia, South Carolina: locally distributed in the northern portions of our range; should be sought in northern Mississippi as well
Plant: terrestrial, 15–60 cm tall
Leaves: 3–5; alternate, somewhat evenly spaced along the stem, spreading; ovate to ovate-elliptic to lance-elliptic, 2.5–12.0 cm wide × 8–20 cm long
Flowers: 1–3(4); sepals and petals spotted, splotched, or marked with brown, chestnut, or reddish-brown spots, rarely appearing as a uniform color; lateral sepals united; petals undulate and spiraled to 10 cm long; lip slipper-shaped, from pale to a deep, rich yellow, less often with scarlet markings within the lip; individual flower size ca. 4.5 × 12.0 cm; lip 2.5–5.4 cm, the opening ovate-oblong at the base of the lip
Habitat: a variety of mesic to calcareous, wet to dry woodlands, streamsides, bogs, and fens
Flowering period: late spring; mid-April–early May in our area

Cypripedium parviflorum var. *pubescens* is the classic yellow lady's-slipper so familiar to many wildflower lovers and gardeners. Although it has declined dramatically in some areas in the past twenty-five years, it still can be found in rich, mesic forests and swamps throughout much of the northern edge of our region. The fact that this is one of the few native orchids than can be cultivated in the garden has led to its decline in the wild. It is not all that unusual to come upon sites where in past years there have been many plants, only to find many holes where they have been dug. Because the plants do grow well under cultivation they should be purchased as propagated plants and there is no need for raping them from the wild!

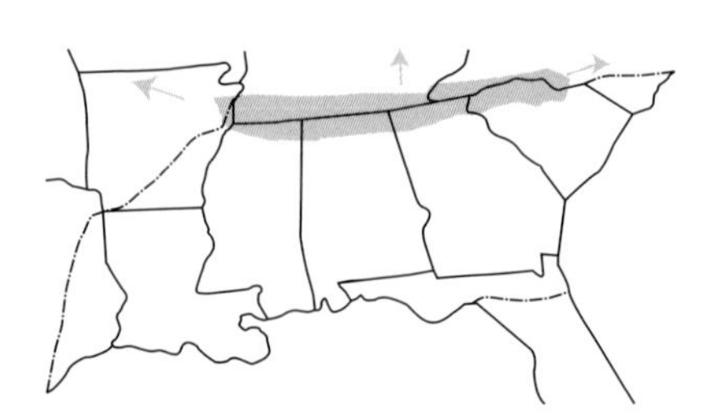

Hybrids:

Three interesting hybrids have often been noted from areas where *Cypripedium parviflorum*, in variety, and *C. candidum* are sympatric. Given the extreme rarity of *C. candidum* and the relative rarity of both varieties of *C. parviflorum* in Alabama the likelihood of the hybrids is very slim. The described hybrids are:

Cypripedium ×*andrewsii* nm *favillianum* (J. T. Curtis) B. Boivin
(*C. candidum* × *C. parviflorum* var. *pubescens*)

Rhodora 34: 100. 1932, as *Cypripedium* ×*favillianum*, type: Wisconsin

Cypripedium ×*andrewsii* nm *landonii* (Garay) Boivin
(*C. candidum* × *C.* ×andrewsii nm *favillianum*)

Canadian Journal of Botany 31: 660. 1953, as *Cypripedium* ×*landonii*, type: Ontario

The latter is a backcross and originally described from southwestern Ontario. The third described hybrid is with *Cypripedium parviflorum* var. *makasin* and of no consequence here. (*Rhodora* 34: 242. 1932, as *Cypripedium* ×*andrewsii*, type: Wisconsin)

Hybrids with *Cypripedium parviflorum* var. *parviflorum* have not been described, although if such do exist they would also fall under the name of *C.* ×*andrewsii*.

Epidendrum

The genus *Epidendrum* contains about 2,000 neotropical species. The seven species found in North America, north of Mexico, are all epiphytes and, with the exception of *E. magnoliae*, all are subtropical. *Epidendrum* (in Florida) is differentiated from *Encyclia* and *Prosthechea*, formerly included in the genus *Epidendrum*, by its sympodial growth and lack of pseudobulbs. Measurements are given for a single growth, but in actuality most species produce clumps of multiple growths.

Epidendrum magnoliae Mühlenberg var. *magnoliae*

green-fly orchis

Florida west to western Louisiana, and north along the Atlantic coast to southeastern North Carolina; Mexico

Louisiana, Alabama, Mississippi, Georgia, Florida, South Carolina, North Carolina: widespread throughout the Gulf and Atlantic Coastal Plains

Plant: epiphytic, to 30 cm tall, often making enormous clumps 50–80 cm across; lithophytic in one site in southern Georgia

Leaves: 2–3; lustrous, dark green, coriaceous; 0.5–1.5 cm wide × 3–10 cm long

Flowers: up to 18, in a loosely flowered, terminal raceme; sepals and petals similar, yellowish-green, lip distinctly 3-lobed and the column with 2 prominent pink calli; individual flower size 1.5–2.5 cm

Habitat: grows primarily on live oak, but also can be found on juniper, magnolia, sweet gum, tupelo, and red maple, often with resurrection fern

Flowering period: August–March, but can flower sporadically throughout the year

Known for more than 150 years as *Epidendrum conopseum* R. Brown *ex* Aiton, Eric Hágsater (2000) recently found that the name *E. magnoliae* predates the name *E. conopseum* by a month. Both were described in 1813. Hágsater writes, "Mühlenberg's *Catalogue* was published in October, whereas Aiton was published in November of the same year. Rules of taxonomic priority require that the earlier name be used instead of the well-known name published by Robert Brown." *Epidendrum magnoliae* is not only the most frequently encountered epiphytic orchid in northern and central Florida, it is also the only epiphytic orchid in the United States that is found outside of Florida. Large plants often occur and their arching flower stems of glowing green flowers can remind one of twinkling holiday lights hanging from the trees.

Epipactis

Epipactis is a cosmopolitan genus of about 25 species, only one of which, *E. gigantea*, is native in North America, well west of the Southeast. Two Eurasian species can also be found in North America, including *E. helleborine* in northern Georgia.

Epipactis helleborine (Linnaeus) Cranz*

broad-leaved helleborine

eastern North America; southeastern California; scattered in western North America; Europe

forma *alba* (Webster) Boivin—white-flowered form
British Orchids, ed. 2. 21. 1898, as *Epipactis latifolia* forma *alba*, type: Wales

forma *luteola* P. M. Brown—yellow-flowered form
North American Native Orchid Journal 4: 318. 1996, type: New Hampshire

forma *monotropoides* (Mousley) Scoggin—albino form
Canadian Field-Naturalist 41: 30. 1927, as *Amesia latifolia* forma *monotropoides*, type: Quebec

forma *variegata* (Webster) Boivin—variegated-leaved form
British Orchids, ed. 2. 22. 1898, as *Epipactis latifolia* forma *variegata*, type: England, Wales

forma *viridens* A. Gray—green-flowered form
Botanical Gazette 4: 202. 1879, type: New York

Georgia: a recent, single record from a rest area along I-75 in northern Georgia
Plant: terrestrial, 10–80 cm tall
Leaves: 3–7; alternate, spreading; lance-elliptic, 2.5–4.0 cm wide × 10–18 cm long
Flowers: 15–50; yellow-green usually suffused with rosy-pink, individual flowers 1–3 cm across
Habitat: highly variable, from shaded calcareous woodlands to front lawns and garden beds and even the crack in a concrete sidewalk!; typically a lime-lover
Flowering period: summer

Epipactis helleborine is a common European species that was first found in North America near Syracuse, New York, in 1878. In the ensuing century-plus it has spread throughout the region and can now be found all the way eastward to down-town Boston, Massachusetts, northward to Nova Scotia and New-foundland, and in recent years westward to California. With the exception of forma *viridens*, the various forms are exceedingly rare and are represented by very few collections.

Although it has not been vouchered for North Carolina, and sparingly for Tennessee, this aggressive visitor is now beginning its spread southward. Plants of *E. helleborine* should be watched for in shaded highway plantings as well as within local parks and woodlands.

forma *viridens*

Eulophia

One of several genera in Florida that have a definite African affinity, *Eulophia* is a pantropical genus of more than 200 species, with only a single species, the wild coco, *E. alta*, found in the Southeast. The flowers bear a distinctive series of keels or crests on the lip. Until recently the genus included species that are now segregated as *Pteroglossaspis* (p. 183). *Eulophia alta* is widespread throughout the West Indies, Central and South America.

Eulophia alta (Linnaeus) Fawcett & Rendle

wild coco

Georgia south to Florida; Mexico, West Indies, Central America, South America, Africa

forma *pallida* P. M. Brown—pale-colored form
North American Native Orchid Journal 1: 132. 1995, type: Florida
forma *pelchatii* P. M. Brown—white/green-flowered form
North American Native Orchid Journal 1: 132. 1995, type: Florida

Florida, Georgia: rare and local in northeastern Florida and adjacent Georgia; at the northern limit of its range
Plant: terrestrial, 50–150 cm tall
Leaves: 4–6; yellow-green, plicate, lanceolate, to 10 cm wide × 100 cm long
Flowers: 20–50; in a tall, loose, many-flowered raceme; sepals and petals similar, lanceolate, highly variable in color from pinks, maroons, and greens to deep rich burgundies or, in the forma *pallida*, pale colored or, in the forma *pelchatii*, light green; the lip usually richer in color or, in the forma *pelchatii*, white, with a pair of prominent crests; individual flower size 3.5–4.5 cm
Habitat: open, lightly wooded swamps and wet woodlands, roadside ditches, and riverbanks
Flowering period: (July) August–October

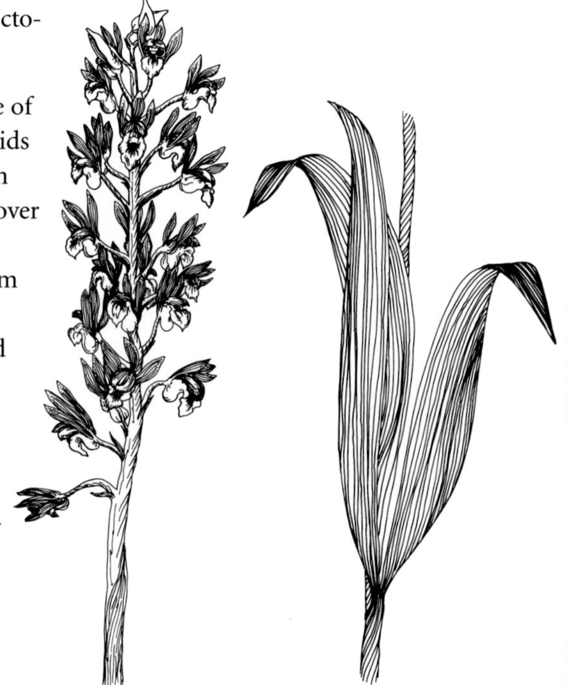

Eulophia alta is the largest and one of the showiest of the terrestrial orchids in the Southeast, is always found in damp to wet ground, and flowers over a long period of time in the late summer and autumn. It grows from a corm and would be an excellent candidate for cultivation if it could be commercially propagated. Flower color is quite variable and both deep, rich burgundies and gentle pale apple-blossom pinks can be found. Because of the relative rarity at the northern limit of its range the two named color forms are less likely to be found.

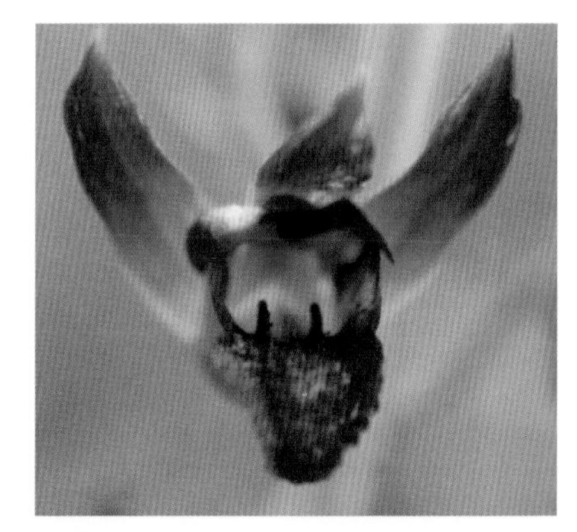

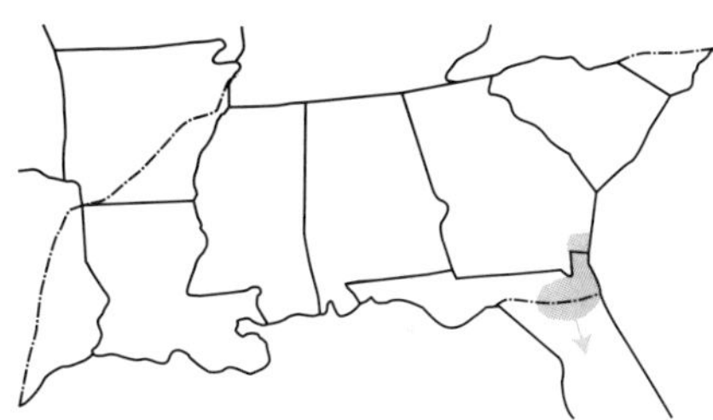

Galearis

Galearis is a small genus of two species with bractless, angled stems and two or three succulent, basal leaves. One species occurs in North America, the other in Asia. Both have showy flowers and have formerly been treated in the genus *Orchis*.

Galearis spectabilis (Linnaeus) Rafinesque

showy orchis

Minnesota east to New Brunswick, south to eastern Oklahoma and Georgia

forma *gordinierii* (House) Whiting & Catling—white-flowered form

New York State Museum Bulletin 243–44: 50. (1921) 1923, as *Galeorchis spectabilis* forma *gordinierii* type: New York

forma *willeyi* (Seymour) P. M. Brown—pink-flowered form

Rhodora 72: 48. 1970, as *Orchis spectabilis* forma *willeyi*, type: Vermont

Alabama, Mississippi, Georgia, South Carolina: local in the northern counties of the Gulf States becoming more frequent in the Appalachians
Plant: terrestrial, 10–15 cm tall
Leaves: 2(4); essentially basal, ovate to oblanceolate, 5–10 cm wide × 6–21 cm long
Flowers: 3–17; in a loose, terminal spike; floral bract conspicuous and often exceeding the flowers; petals and sepals lavender-purple, lip white; in the forma *gordinierii* the flowers entirely white or, in the forma *willeyi*, the flowers entirely purple; individual flower size ca. 3.5 × 4.5 cm
Habitat: rich mesic or calcareous woodlands, often on slopes and streamsides
Flowering period: spring; late March–May

Galearis is an orchid that is typical of the rich mesic forest of the Appalachians and upper Piedmont. The species reaches the southern limit of its range in the northern counties of our area. The fleshy, almost succulent, leaves are distinctive and the plants are often the favorite rooting area for skunks, who like to feast on the tasty roots. Many a time it has been assumed that orchid-hungry thieves have dug the plants when it has been our striped friends!

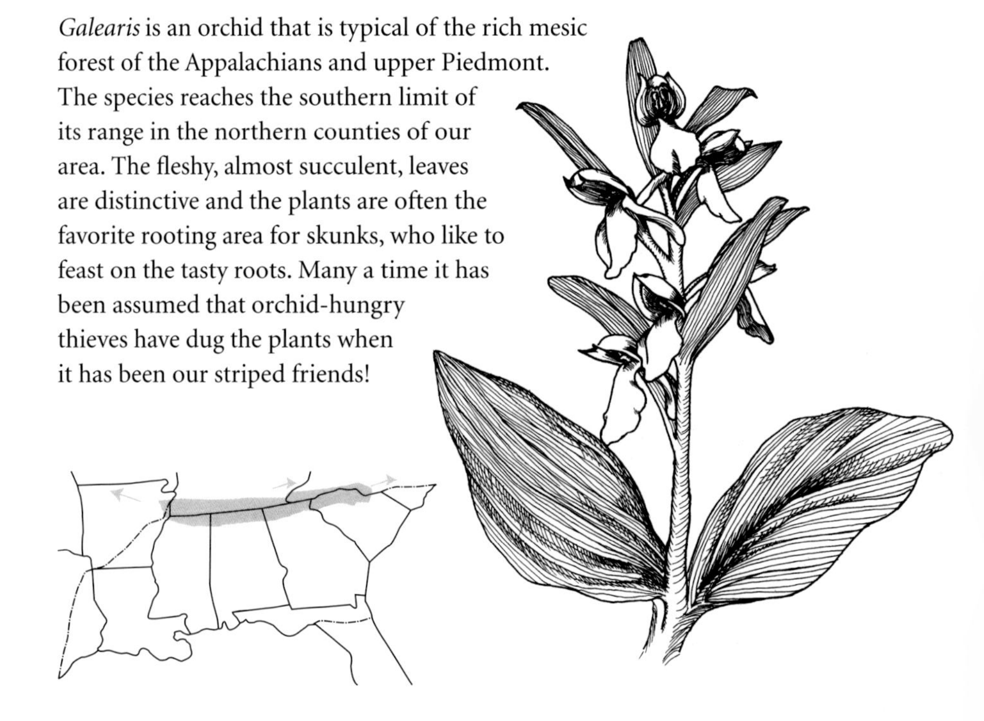

forma *willeyi*

forma *gordinierii*

Goodyera

Goodyera is a terrestrial genus that is widespread throughout the world and known for its beautifully marked and reticulated leaves, often earning the group the name of "jewel orchids." In the United States and Canada we have four species, all primarily northern or higher elevation in distribution. Only one species, the **downy rattlesnake orchis**, *G. pubescens*, is found in our area, where it reaches the southern limit of its range. *Goodyera repens*, the **lesser rattlesnake orchis**, is found in the adjacent mountains of North Carolina. See page 245 for additional details.

Goodyera pubescens (Willdenow) R. Brown

downy rattlesnake orchis

Ontario east to Nova Scotia, south to Arkansas and Florida
Alabama, Arkansas, Mississippi, Georgia, Florida, South Carolina, North Carolina: local in the northern and central counties and becomes very rare southward
Plant: terrestrial, 20–30 cm tall
Leaves: 4–6; in a basal rosette, green with white reticulations, lanceolate, 1.0–1.5 cm wide × 5–10 cm long
Flowers: 20–50+; in a densely flowered terminal spike; white, copiously pubescent; individual flower size 3 × 4 mm
Habitat: mixed and deciduous woodlands
Flowering period: July–August (November)

It is interesting that some of the rarest orchids at the southern limit of our range are also some of the most frequently encountered species as one travels northward. The downy rattlesnake orchis falls into this category, along with the little club-spur orchis and large whorled pogonia. In the Southeast, like the latter two, this species is restricted to the more northern affinity soils and woodlands, and becomes increasingly frequent throughout the central and northeastern United States. It has the most handsomely marked foliage of any of our native orchids and also has the added feature of being evergreen.

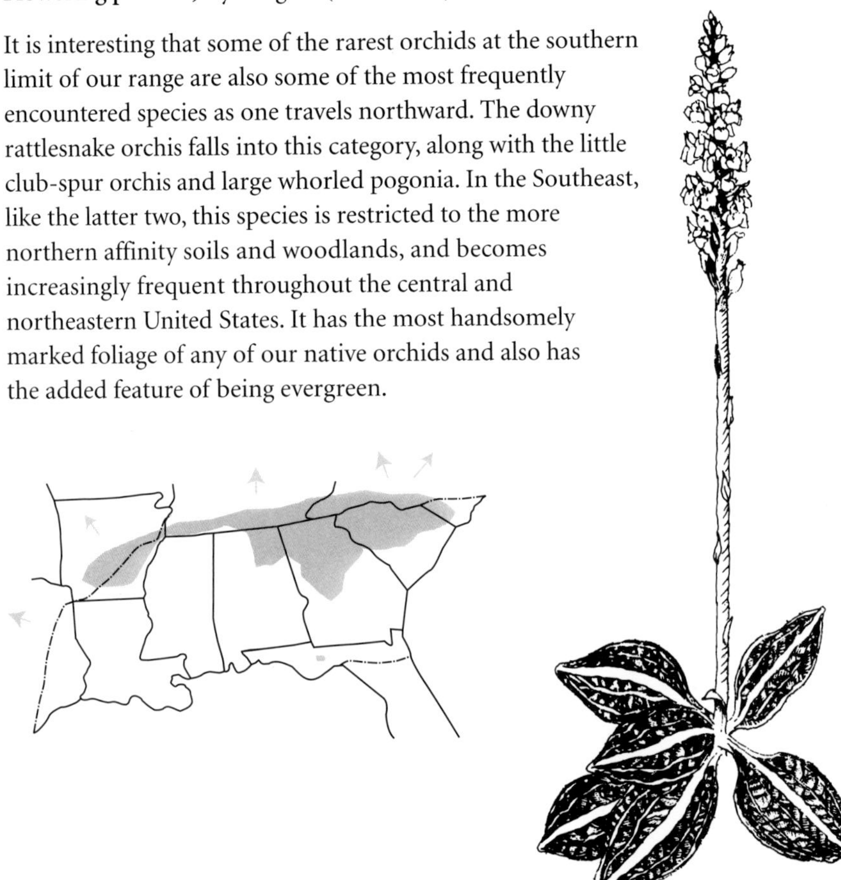

Gymnadeniopsis

Gymnadeniopsis was formerly placed within the genera *Habenaria* and *Platanthera*. The three species that Rydberg used to comprise the genus *Gymnadeniopsis* have recently been reassessed (Sheviak FNA, 2002; Brown, 2003). Several differences are present that render them distinctive. The presence of tubers on the roots as well as small tubercles on the column are two of the major differences that separate *Gymnadeniopsis* from the other related genera.

1a flowers white; non-resupinate (lip uppermost) . . . **snowy orchis**, *G. nivea*
1b flowers otherwise . . . 2

2a flowers yellow; margin erose . . . **yellow fringeless orchis**, *G. integra*
2b flowers green to straw-colored to nearly white; spur swollen at the tip . . . **little club-spur orchis**, *G. clavellata*

Gymnadeniopsis clavellata (Michaux) Rydberg var. *clavellata*

little club-spur orchis

Wisconsin east to Maine, south to Texas and Georgia

forma *slaughteri* (P. M. Brown) P. M. Brown—white-flowered form

North American Native Orchid Journal 1(3): 200. 1995, as *Platanthera clavellata* var. *clavellata* forma *slaughteri*, type: Arkansas

Texas, Louisiana, Arkansas, Alabama, Mississippi, Georgia, Florida, South Carolina, North Carolina: not at all uncommon north of the Gulf Coastal Plain, this species becomes increasingly frequent northward

Plant: terrestrial, 15–35 cm tall

Leaves: 2; cauline, ovate-lanceolate, 1–2 cm wide × 5–15 cm long

Flowers: 3–15; often arranged in a short, dense, terminal raceme, flowers usually twisted to one side; sepals ovate, petals linear, enclosed within the sepals forming a hood; lip oblong, the apex obscurely 3-lobed; perianth yellow-green or, in the forma *slaughteri*, white; individual flower size 0.5 cm, not including the 1 cm spur, the small tip swollen (clavellate)

Habitat: damp woods, streamsides, open, wet ditches

Flowering period: June–August

The little club-spur orchis is the second species we have of a common northern orchid that reaches its southern limit in our region. In Florida it is confined to a single location. The small, pale greenish flowers are very different from any other orchid we have. They also hold themselves at curious angles on the stem. The distinctive spur, with its swollen tip, is what gives this plant its common name. *Gymnadeniopsis clavellata* var. *ophioglossoides* is found primarily in the northern and higher elevation areas of northeastern North America.

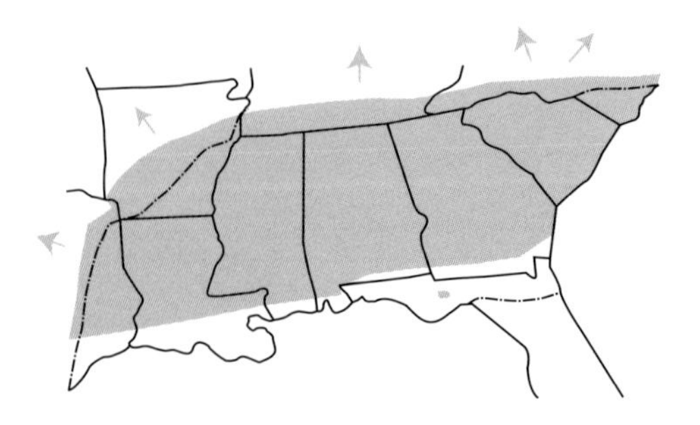

forma *slaughteri*

Gymnadeniopsis integra (Nuttall) Rydberg

yellow fringeless orchis

New Jersey south to Florida and west to Texas
Texas, Louisiana, Alabama, Mississippi, Georgia, Florida, South Carolina, North Carolina: formerly widespread throughout most of the Coastal Plain region, there are few remaining extant sites
Globally Threatened
Plant: terrestrial, to 60 cm tall
Leaves: 1–2(3); cauline, lanceolate, keeled, conduplicate, 1–3 cm wide × 5–20 cm long, rapidly reduced to bracts within the inflorescence
Flowers: 30–65; arranged in a densely flowered terminal raceme; sepals ovate, petals oblong-ovate, enclosed within the dorsal sepal forming a hood; lip ovate, the margin erose; perianth bright yellow-orange; individual flower size 5–6 mm, not including the 6 mm spur
Habitat: open wet meadows, seeps, damp pine flatwoods
Flowering period: late July–September

Although the yellow fringeless orchis ranges from southern New Jersey to Texas, it is rapidly becoming one of the rarer orchids to be found in North America. Habitat destruction, specifically the draining of wetlands for agriculture and business, is the primary cause of the decline. There are still a few excellent scattered sites in the Florida panhandle within the Apalachicola National Forest. Otherwise the sites are very local and do not flower reliably each year. The plants of *Gymnadeniopsis integra* are not difficult to distinguish from the similarly colored fringed orchises (*Platanthera* spp.), as they are considerably smaller and more slender, with a more compact, conical raceme of flowers, and lack the distinctive fringe on the margin of the lip.

Gymnadeniopsis nivea (Nuttall) Rydberg

snowy orchis

southern New Jersey south to Florida and west to Texas
Texas, Louisiana, Arkansas, Alabama, Mississippi, Georgia, Florida, South Carolina, North Carolina: rare and local but in a few places locally abundant, especially in southern Georgia
Plant: terrestrial, 20–60 cm tall
Leaves: 2–3; cauline, lanceolate, keeled, conduplicate, 1–3 cm wide × 5–25 cm long
Flowers: 20–50; non-resupinate, arranged in a densely flowered terminal raceme; sepals and petals oblong-ovate; lip uppermost, linear-elliptic and bent backwards midway; perianth stark, icy white; individual flower size 8–10 mm, not including the slender 1.5 cm spur
Habitat: open wet meadows, prairies, seeps, damp pine flatwoods
Flowering period: late May–July

Similar in habit and habitat to the **yellow fringeless orchis,** *Gymnadeniopsis integra*, the snowy orchis is far more widespread and occurs in scattered locales throughout the Coastal Plain. *Gymnadeniopsis nivea* is not quite as rare as *G. integra* and in southern Georgia large colonies can often be found in early June along the broad, mown roadsides of US 301. Recently burned pine flatwoods are also an excellent habitat to look for this orchid. It has the added feature of being deliciously fragrant, and the uppermost lip is unique among the species of *Gymnadeniopsis, Habenaria,* and *Platanthera* found in the southeastern United States.

Hybrid:

With revalidation of the genus *Gymnadeniopsis*, the previously described hybrid between *Platanthera clavellata* and *Platanthera blephariglottis*—*Platanthera* ×*vossii* Case—needed a new combination and a new hybrid genus. The resulting combination was:

X*Platanthopsis vossii* (Case) P. M. Brown
North American Native Orchid Journal 8:39–40. 2003, type: Michigan.

Because of the range of the two parents in the Southeast, the likelihood of the hybrid is exceedingly rare.

Habenaria

The genus *Habenaria* is pantropical and subtropical and consists of about 600 species. It reaches the northern limit of its range in North America in the southeastern United States. In its broadest sense, the genus often includes those species that are found in *Platanthera*, *Coeloglossum*, *Gymnadeniopsis*, and *Pseudorchis*. As treated here, in the narrow sense, it contains five species, three found within the southeastern United States and the other two confined to central and southern Florida. All three of the species in our area may form large colonies of distinctive basal rosettes.

1a lip and/or petals divided into linear, threadlike segments . . . 2
1b lip and/or petals merely toothed . . . **toothed rein orchis,** *H. odontopetala*

2a spur equal to the ovary; plants of wet habitats . . . **water spider orchis,** *H. repens*
2b spur distinctly longer than the ovary . . . **Michaux's orchis,** *H. quinqueseta*

Habenaria odontopetala Reichenbach *f.*

toothed rein orchis

Florida; Mexico, West Indies, Central America
forma *heatonii* P. M. Brown—albino form
North American Native Orchid Journal 7(1): 93–94. 2001, type: Florida

Florida: reaches the northern limit of its range in northeastern Florida not many miles from the Georgia border
Plant: terrestrial, up to 1 m tall
Leaves: 5–12; glossy green, elliptic, to 3–5 cm wide × 20 cm long, or in the forma *heatonii*, the entire plant white, lacking chlorophyll; the leaves gradually reduced in size and passing to bracts within the inflorescence
Flowers: 10–60; in an often densely flowered terminal raceme; sepals green, ovate to oblong; petals yellow-green with 2 obscure teeth or, in the forma *heatonii*, white; lip with 3 divisions, the central one being noticeably longer than either the petals or sepals; spur ca. 2.5 cm long; individual flower size ca. 2.0 cm × 2.5 cm
Habitat: rich, damp hardwood hammocks
Flowering period: October–November

The toothed rein orchis, one of the most common orchids in central and southern Florida, reaches the northern limit of its range in Jacksonville, Duval County, Florida. Although the green flowers are somewhat inconspicuous, the verdant basal rosettes and large, bold plants are unmistakable in the autumn woodlands. The flowers have a distinctly unpleasant odor.

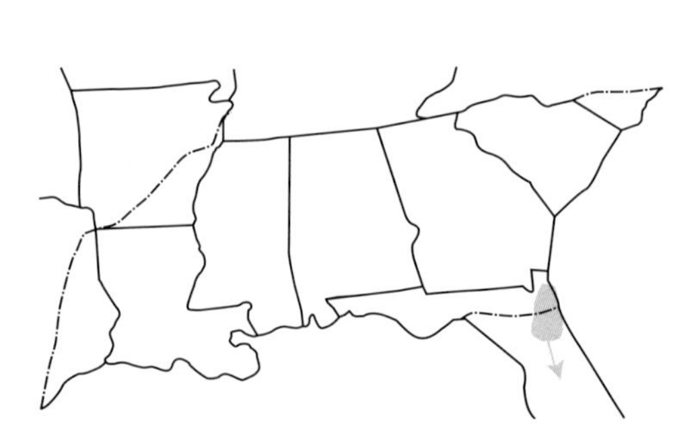

Habenaria quinqueseta (Michaux) Eaton

Michaux's orchis

South Carolina south to Florida, and west to Texas; Mexico, West Indies, Central America

Texas, Louisiana, Alabama, Mississippi, Georgia, Florida, South Carolina: very rare and local in all but Florida; presumed extirpated in Texas and South Carolina

Plant: terrestrial, up to 30(50) cm tall

Leaves: 3–7; glossy bright green, elliptic, to 4–6 cm wide × 20 cm long; gradually reduced in size and passing to bracts within the inflorescence

Flowers: 15–25; in a loose raceme; sepals light green with dark green stripes, ovate to oblong; petals white with 2 linear divisions; lip white with 3 divisions; spur 5–8 cm long, but local clones may have longer spurs; individual flower size ca. 4 × 4 cm, not including the spur

Habitat: rich, moist hardwood hammocks, pine flatwoods, roadside ditches

Flowering period: August–September

Michaux's orchis is widespread and locally common throughout much of northern Florida. Elsewhere it is one of the rarest orchids to be found. This is primarily a plant of damp pinelands and hedgerows. The greenish-white flowers are produced on a spike to 30+ cm tall and with up to 14 flowers, but more often with only 6–8 flowers. The spur typically is shorter than 5 cm. Large colonies of nonflowering plants are often encountered, especially in open pine flatwoods. The flowers of *Habenaria quinqueseta* have a very square aspect to them whereas those of the more southern, and closely allied, *H. macroceratitis* have a very rectangular aspect.

Habenaria repens Nuttall

water-spider orchis

North Carolina south to Florida, and west to southeastern Arkansas and Texas; Mexico, West Indies, Central America

Texas, Arkansas, Louisiana, Alabama, Mississippi, Georgia, Florida, South Carolina: widespread throughout the southeastern United States, primarily on the Coastal Plain

Plant: terrestrial or aquatic, up to 50 cm tall

Leaves: 3–8; yellow-green, linear-lanceolate 1.0–2.5 cm wide × 3–20 cm long; rapidly reduced in size and passing to bracts within the inflorescence

Flowers: 10–50; in a densely flowered terminal raceme; sepals light green, ovate to oblong; petals greenish-white, with 2 divisions; lip with 3 divisions, the central being shorter than the lateral divisions; spur slender ca. 1.3 cm long; individual flower size ca. 2 cm × 2 cm, not including the spur

Habitat: open shorelines, wet ditches, stagnant pools

Flowering period: throughout the year

The water-spider orchis is one of the few truly aquatic orchids. Masses of several hundred floating plants can often be found, and it also frequently colonizes wet roadside ditches and canals. This species produces fewer sterile colonies than other species of *Habenaria*, and it is also the most wide-ranging of the *Habenaria* species in the United States.

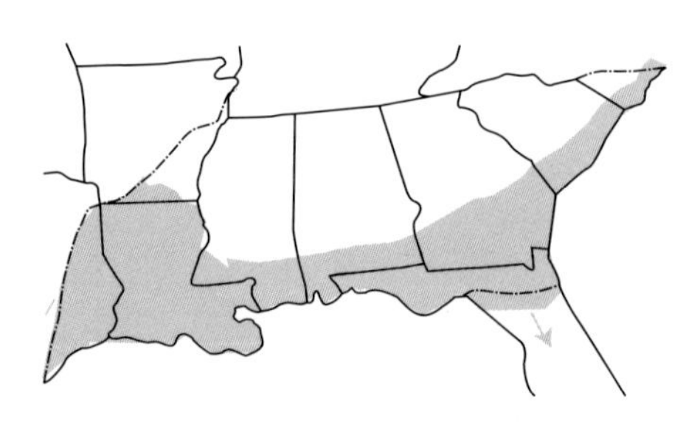

Hexalectris

A genus of seven species that is found primarily in the southern United States and Mexico, *Hexalectris* is similar in appearance to the other mycotrophic genus, *Corallorhiza*. Although not that closely related, they both have colorful flowers terminating a leafless stem that lacks all chlorophyll. The flowers on *Hexalectris* are much larger than those of *Corallorhiza*, and intricately more beautiful. The only species to occur in the Southeast, the crested coralroot, *H. spicata*, is also the only species to range east of Texas from the southwestern United States and adjacent Mexico.

Hexalectris spicata (Walter) Barnhardt var. *spicata*

crested coralroot

Arizona, Missouri; southern Illinois east to Maryland, south to Florida, and west to Texas; Mexico

forma *albolabia* P. M. Brown—white-lipped form

North American Native Orchid Journal 1(1): 10. 1995, type: Florida

Texas, Louisiana, Arkansas, Alabama, Mississippi, Georgia, Florida, South Carolina, North Carolina: widespread, although not necessarily common
Plant: terrestrial, mycotrophic, 10–80 cm tall; stems yellow-brown to deep purple
Leaves: lacking
Flowers: 5–25; sepals and petals brown to yellow with purple striations; lip pale yellow with purple stripes (crests), 3-lobed, the lateral lobes incurved or, in the forma *albolabia,* the lip pure white with pale yellow striations and the petals and sepals mahogany; individual flower size 2.5–4.0 cm
Habitat: dry, open hardwood forest especially under live oaks
Flowering period: June–August

The crested coralroot is by far the handsomest of all the terrestrial leafless orchids, and one of the most beautiful of the summer-flowering orchids in the southern United States. One of several species of entirely mycotrophic orchids to be found in the Southeast, the plants lack all traces of chlorophyll and therefore often blend in with their surroundings. From a distance *Hexalectris spicata* may appear to be just a dead stick, but upon closer examination the striking and colorful flowers reveal an intricate pattern of crests upon the lip. Plants have a preference for live oak woodlands and the number of plants, depending on the rainfall in a given season, may vary greatly from year to year. *Hexalectris spicata* var. *arizonica* occurs from central Texas westward.

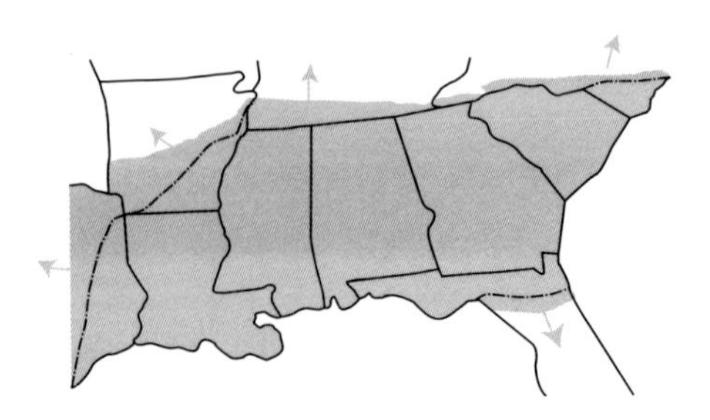

Isotria

The genus *Isotria* consists of only two species, both found in the eastern United States and adjacent Canada. They are related to the genera *Pogonia* and, more distantly, *Triphora*. Early in their history they were placed in the genus *Pogonia*. We have both species in the Southeast.

1a sepals greenish-yellow; one and one-half times as long as the petals or shorter than the petals . . . **small whorled pogonia**, *I. medeoloides*

1b sepals purple; two or more times as long as the petals . . . **large whorled pogonia**, *I. verticillata*

Isotria medeoloides (Pursh) Rafinesque

small whorled pogonia

Michigan east to Maine, south to Missouri, Georgia, and South Carolina
FEDERALLY LISTED AS THREATENED
Georgia, South Carolina, North Carolina: very rare in all states in which it occurs, often no more than a few plants
Plant: terrestrial, mature plants up to 15 cm tall, shorter (8–12 cm) in flower
Leaves: 5 or 6; in a whorl at the top of the stem, up to 1 cm wide × 5 cm long
Flowers: 1 or 2; sepals and petals greenish-yellow, wide spreading; lip white; individual flowers ca. 2–3 cm across
Habitat: various wooded habitats; favoring beech, mixed pines, etc.; often near seasonal runoffs
Flowering period: April–May

Isotria medeoloides was one of the first orchids to be listed by the federal government under the Endangered Species Act. Although very rare, *I. medeoloides* is often known in most states from a single station, or even a single plant in southeastern North America. Further north, in New Hampshire and western Maine, the small whorled pogonia can be locally abundant. Both states have sites in excess of several thousand plants. This is an excellent example of a species originally thought to be one of the very rarest in North America. With the advent of more people, both professional and amateur, searching, more plants in many new sites have been documented.

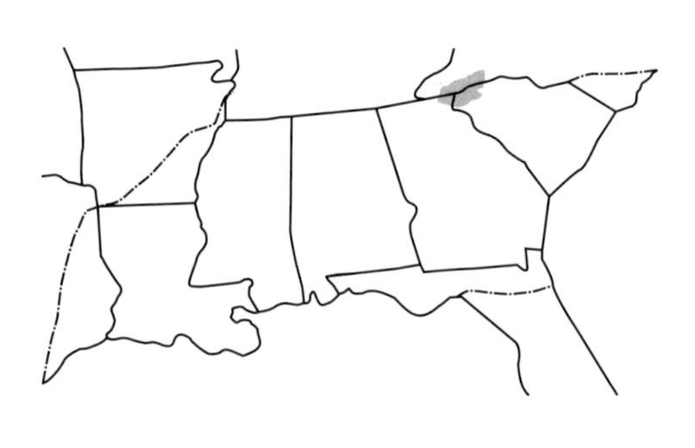

Isotria verticillata (Willdenow) Rafinesque

large whorled pogonia

Michigan east to Maine, south to Texas and Florida
Texas, Louisiana, Arkansas, Alabama, Mississippi, Georgia, Florida, South Carolina, North Carolina: locally common throughout our region, especially north of Florida
Plant: terrestrial, mature plants up to 40 cm tall, shorter (to 20 cm) in flower
Leaves: 5 or 6; in a whorl at the top of the stem, up to 3 cm wide × 9 cm long
Flowers: 1, rarely 2; sepals purplish, wide-spreading, slender, and spidery; petals pale yellow, ovate, arched over the column; lip white, edged in purple with a fleshy, yellow central ridge; individual flowers ca. 10 cm across
Habitat: deciduous forests
Flowering period: late March to May, usually before the trees leaf out

Isotria verticillata is the third of the three species that are very common northward, but very rare at the southern limit of their range. *Isotria verticillata* along with *Goodyera pubescens* and *Gymnadeniopsis clavellata* form this trio, and all have a preference for northern affinity habitats of cooler and richer woodlands. Although not always sympatric here, they often grow in the same community northward. The fanciful flowers of *Isotria* are unmistakable and the plants have an unusual habit (as does the only other species in the genus, *I. medeoloides*) of nearly doubling in both leaf surface and height after flowering.

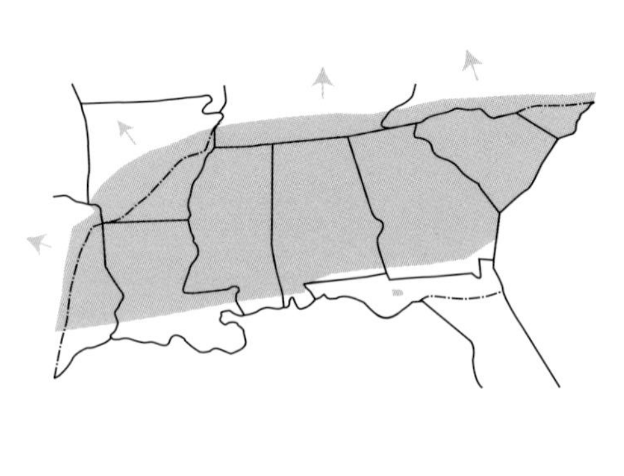

Liparis

Liparis is a cosmopolitan genus of more than 200 species occurring in the widest variety of habitats throughout the world. All members of the genus are terrestrial or semi-epiphytic and have swollen bases to the leaves that form pseudobulb-like structures. These structures are not unlike those of the genus *Malaxis*; in the subtropical and tropical species they are more evident and usually above ground, whereas in the temperate and more northerly species the pseudobulb-like structures are within the ground. Three species occur in the United States and Canada, but only two within the Southeast.

1a flowers yellow-green; plants usually of moist areas, borrow pits, grassy fens, etc. . . . **Löesel's twayblade,** *L. loeselii*

1b flowers chocolate-purple; plants of rich woodlands . . . **lily-leaved twayblade,** *L. liliifolia*

Liparis liliifolia (Linnaeus) Richard *ex* Lindley

lily-leaved twayblade

Minnesota and Ontario east to New Hampshire, south to Oklahoma and Georgia
forma *viridiflora* Wadmond—green-flowered form
Rhodora 34: 18. 1932, type: Wisconsin

Alabama, Mississippi, Georgia, South Carolina: scattered populations that can be locally abundant in the northern counties and rich woodlands of the Piedmont
Plant: terrestrial, 9–30 cm tall
Leaves: 2; basal, green, strongly keeled, ovate 4 cm wide × 6–8 cm long
Flowers: 5–78; in a terminal raceme; sepals purple, ovate; petals and sepals green, slender and threadlike; lip chocolate-purple, broadly ovate or, in the forma *viridiflora*, flowers entirely green; individual flower size 1.2–2.4 cm
Habitat: rich, damp woodlands, mesic forest, shaded banks and roadsides often occurring in calcareous soils
Flowering period: late spring; May–June

Liparis liliifolia, a broad, handsome orchid of rich mesic forests and streamsides, is one of the prime components of the southern Appalachian woodlands. It is widespread in its distribution though usually absent from the Coastal Plain, and is often found growing with several other orchids including the **putty-root**, *Aplectrum hyemale*, and **showy orchis**, *Galearis spectabilis*.

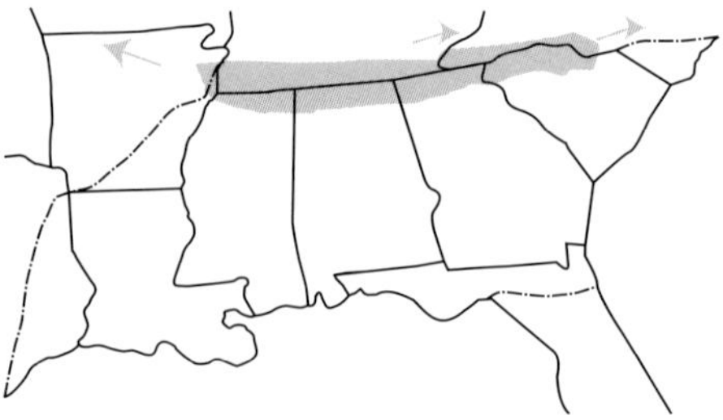

forma *viridiflora*

Liparis loeselii (Linnaeus) Richard

Löesel's twayblade, fen orchis

British Columbia east to Nova Scotia, south to Arkansas and Mississippi, and the southern Appalachian Mountains; Europe
Alabama, Mississippi: a northern species that just reaches the southern limit of its range in the northern counties of these two states; it should be actively sought in northern Georgia and adjacent South Carolina
Plant: terrestrial, 4–20 cm tall
Leaves: 2; basal, pale green, strongly keeled; oblanceolate, 2–3 cm wide × 4–6 cm long
Flowers: 5–15; in a terminal raceme; sepals, petals, slender and threadlike; lip, broadly ovate; watery-green; individual flower size 0.5–1.0 cm
Habitat: damp gravels; bogs; ditches, seepages, shaded banks and roadsides often in calcareous soils
Flowering period: late spring; May–June

The fen orchis is one of the few species that the eastern United States shares with northern Europe. And as rare as it is in Europe, it can be common north of our region. Because of its translucent coloring it is easily overlooked. Just north of our area, along the southern Blue Ridge Parkway, it occurs with *Liparis liliifolia* and produces the very rare hybrid *L.* ×*jonesii*, Jones' hybrid twayblade. This unusual hybrid has not been documented elsewhere, although both parent species occur in Alabama and Mississippi, as well as many other places in the eastern United States.

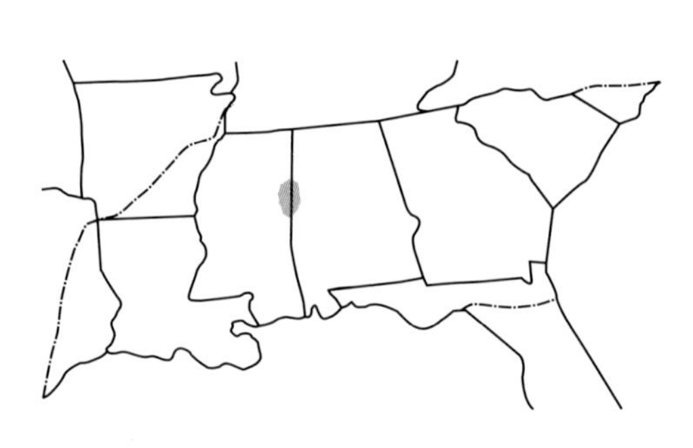

Hybrid:
Liparis ×jonesii S. Bentley
Jones' hybrid twayblade
(*L. liliifolia* × *L. loeselii*)

> *Native Orchids of the Southern Appalachian Mountains* 1999. p. 138–39, type: North Carolina

Jones' hybrid twayblade is very rare and currently known only from the type locality along the Blue Ridge Parkway in southwestern North Carolina, just north of our limits for the Southeast.

Listera

The genus *Listera* is composed of 25 species that occur in the cooler climes of both the Northern and Southern Hemispheres. Eight species in the genus grow in the United States and Canada, one of which, *L. ovata*, is a very common species in Europe that has become naturalized in southern Ontario. We have only two species in the Southeast.

1a lip deeply cleft to more than half its length . . . **southern twayblade,** *L. australis*

1b lip shallowly cleft to less than half its length . . . **Small's twayblade,** *L. smallii*

Listera australis Lindley

southern twayblade

Quebec to Nova Scotia, and south to Florida and Texas

forma *scottii* P. M. Brown—many-leaved form

North American Native Orchid Journal 6(1): 63–64. 2000, type: Florida

forma *trifolia* P. M. Brown—3-leaved form

North American Native Orchid Journal 1(1): 11. 1995, type: Vermont

forma *viridis* P. M. Brown—green-flowered form

North American Native Orchid Journal 6(1): 63–64. 2000, type: Florida

Texas, Louisiana, Arkansas, Alabama, Mississippi, Georgia, Florida, South Carolina, North Carolina: locally common throughout
Plant: terrestrial in damp soils, up to 35 cm tall
Leaves: 2; opposite, midway on the stem, green, usually flushed with red, ovate 2.0 cm wide × 3.5 cm long or, in the forma *scottii*, leaves several scattered along the stem or, in the forma *trifolia*, leaves 3 in a whorl
Flowers: 5–40; in a terminal raceme; sepals purple, ovate; petals purple, narrowly spatulate, recurved; lip purple, linear, split into 2 slender filaments or, in the forma *viridis*, flowers entirely green; individual flower size 6–10 mm
Habitat: rich, damp woodlands, often in sphagnum moss
Flowering period: late December–March

A spring ephemeral, the southern twayblade appears quickly in January (in Florida) and throughout February and March in the north-central and northerncounties of the Southeast, having a preference for damp, often seasonally flooded, deciduous woodlands. In many sites, it prefers the presence of sphagnum moss. Although most populations consist of fewer than a dozen plants, occasional sites may have several thousand individuals and contain an amazing degree of variation with all described forma present. The plants at many of these larger sites tend to be more robust than elsewhere and flower over a very long period of time, forming dense, lush clumps. The species sets seed and senesces very quickly so that a month after flowering there usually is no sign of the plants until next season.

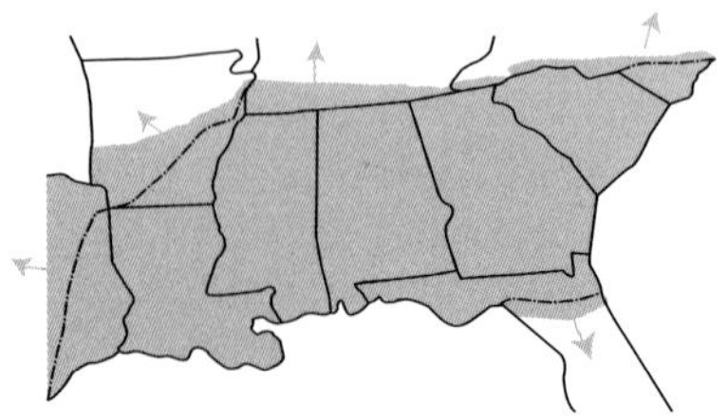

forma *trifolia*

forma *scottii*

forma *viridis*

Listera smallii Wiegand

Small's twayblade

northern New Jersey; central Pennsylvania south to northern Georgia
forma *variegata* P. M. Brown—variegated-leaved form
North American Native Orchid Journal 1(4): 289. 1995, type: West Virginia

Globally Threatened
Georgia, South Carolina: rare and local in the Appalachian Mountains
Plant: terrestrial in damp soils, up to 25 cm tall
Leaves: 2; opposite midway on the stem, reniform to ovate, 2.0 cm wide × 3.5 cm long, green or, in the forma *variegata*, with white markings
Flowers: 4–12; in a terminal raceme; greenish-yellow to beige; lip broadened at the summit and notched; individual flower size 6–12 mm
Habitat: rich, damp woodlands, and along stream runs, usually under *Rhododendron maximum*
Flowering period: late June–July

This is the rarest twayblade in North America and is known only from deep within the Appalachian Mountains. The distribution is very specific from south-central Pennsylvania, near State College, south to northeastern Georgia, with a disjunct site in the northern mountains of New Jersey. The neat little broad-lipped flowers vary in color from translucent green to beige. Plants are often tucked up under the lower branches of the great laurel, *Rhododendron maximum*, and many plants most likely occur within the "rhododendron hells" in the southern Appalachians.

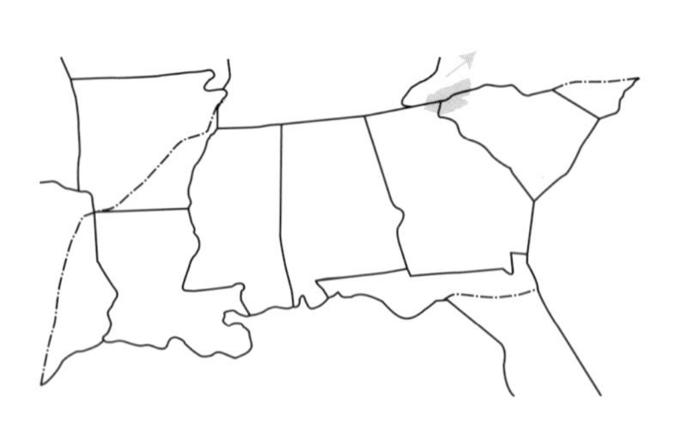

Malaxis

The genus *Malaxis* is a cosmopolitan one of about 300 species. Eleven species are found in the United States and Canada, three of which occur in the Southeast. All species have a pseudobulbous stem that is more evident in the subtropical and tropical species. In the temperate species it appears more cormlike. The genus possesses some of the smallest flowers in the terrestrial Orchidaceae, many not greater than a few millimeters in any dimension.

1a inflorescence a slender spike . . . **Florida adder's-mouth**, *M. spicata*
1b inflorescence otherwise . . . 2

2a inflorescence an elongated cluster or raceme; lower flowers persisting after anthesis . . . **Bayard's adder's-mouth**, *M. bayardii*
2b inflorescence a flat-topped cluster; lower flowers withering after anthesis . . . **green adder's-mouth**, *M. unifolia*

Malaxis bayardii Fernald

Bayard's adder's-mouth

Nova Scotia; Massachusetts south to South Carolina, and west to Ohio
South Carolina: very rare in both the mountains and Piedmont
Globally Threatened
Plant: terrestrial in dry, open ground; 4–18+ cm tall, stem swollen at the base into a (pseudo)bulb
Leaves: 1; ovate, keeled, to 6 cm wide × 9 cm long, midway on the stem
Flowers: 35–150; arranged in an elongated raceme; sepals oblanceolate, yellow-green; petals linear and positioned behind the flower; lip yellow-green, broadly ovate to cordate, with prominent auricles at the base and bidentate at the summit; individual flower size 2–4 mm, persisting after anthesis
Habitat: dry open woodlands, pine flatwoods, roadsides, rocky barrens
Flowering period: June

Malaxis bayardii is perceived to be very rare, and is perhaps overlooked. Plants of *M. unifolia* found growing in atypical dry habitats should be carefully examined to see if they are *M. bayardii*, as the overall distribution is not well known. Recently revalidated as a species (Catling, 1991), this diminutive green orchid is currently known globally from only six sites, three of which are within our region. Its inclusion in this work is based on the discovery of several herbarium specimens from South Carolina and a possible extant site near Greenville. Historically it was vouchered from North Carolina, Virginia, New Jersey, and extensively from Pennsylvania, as well as New England and New York. The largest site of ca. 400 individuals occurs on an abandoned woodroad and the adjacent field sides on Cape Cod, Massachusetts. Its similarity to *M. unifolia* makes for some problems in identification, but Catling's article goes into great detail and supports all identification criteria. Unless more sites can be found this is an excellent candidate for federal listing as threatened or endangered.

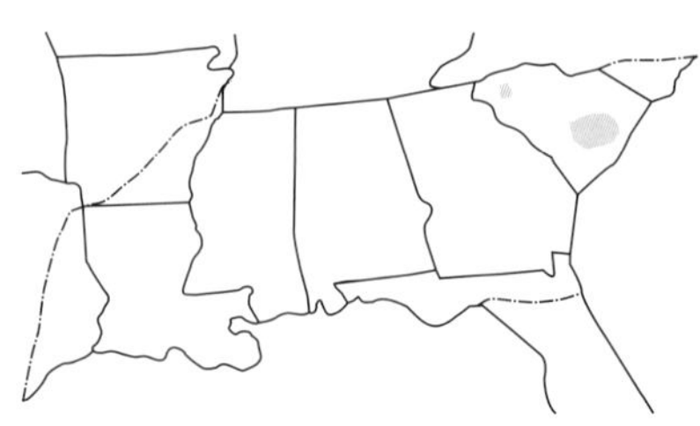

Malaxis spicata Swartz

Florida adder's-mouth

Virginia south to Florida; the Bahamas, West Indies
forma *trifoliata* P. M. Brown—3-leaved form
North American Native Orchid Journal 9: 34. 2003, type: Florida

Georgia, Florida, South Carolina, North Carolina: widespread and not infrequent in wooded swamps along the Coastal Plain
Plant: terrestrial or semi-epiphytic in damp soils, rotted logs, and old tree stumps; 8–56 cm tall, stem swollen at the base into a pseudobulb
Leaves: (1)2 or, in the forma *trifoliata,* 3; basal, subopposite, ovate, keeled, 1–5 cm wide × 2–10 cm long
Flowers: 5–115; doubly resupinate and arranged in a spike; sepals oblanceolate, brown to green; petals similar, linear, and positioned behind the flower; lip orange, vermilion, brown, or green, uppermost, broadly ovate to cordate, with extended auricles at the base; individual flower size 3–5 mm
Habitat: rich, damp woodlands, riverbanks and floodplains, floating logs, tree stumps and bases
Flowering period: August–September

The Florida adder's-mouth is widespread, if not locally common. It grows readily along the floodplains and wet woods throughout the area. Plants tend to be small, usually less than 20 cm, in the cooler northern counties, but southward it is not unusual to find plants well over 30 cm in height with more than 100 minute yellow, orange, and green flowers. The leaves arise from a pea-sized pseudobulb that is nestled in moist soils or occasionally in older, rotting stumps or tree cavities. Plants flower successively over a very long period of time, often bearing flowers for more than three months.

Malaxis unifolia Michaux

green adder's-mouth

Manitoba east to Newfoundland, south to Texas and Florida; Mexico

forma *bifolia* (Mousley) Fernald—2-leaved form

Orchid Review 35: 163. 1927, type: Quebec

forma *variegata* Mousley—variegated-leaf form

Orchid Review 35: 164. 1927, type: Quebec

Texas, Louisiana, Arkansas, Alabama, Mississippi, Georgia, Florida, South Carolina, North Carolina: rare and local throughout its range; easily overlooked
Plant: terrestrial in dry to damp mixed woodlands; 8–25+ cm tall, stem swollen at the base into a (pseudo)bulb
Leaves: 1, or 2 in the forma *bifolia*; ovate, keeled, to 6 cm wide × 9 cm long, midway on the stem; green or, in the forma *variegata*, with white markings
Flowers: 5–80+; arranged in a compact raceme, elongating as flowering progresses; sepals oblanceolate, green; petals linear and positioned behind the flower; lip green, broadly ovate to cordate, with extended auricles at the base and bidentate at the summit; individual flower size 2–4 mm
Habitat: dry to damp woodlands, mesic pine flatwoods
Flowering period: February–May (June)

The green adder's-mouth is as rare in many areas of the southern Atlantic Coastal Plain as the Florida adder's-mouth may be frequent. They have very little in common other than their morphology. Whereas the Florida adder's-mouth is a moisture lover, the green adder's-mouth prefers dry woodlands and mesic pine flatwoods. All but one of the few populations in the Southeast is small, with fewer than fifty plants. Large plants are not uncommon and they, like most members of the genus, bear up to 100 flowers and present them over a long period of time—up to two months.

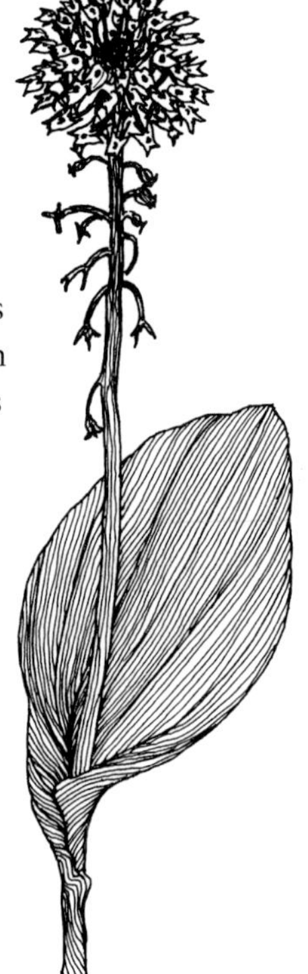

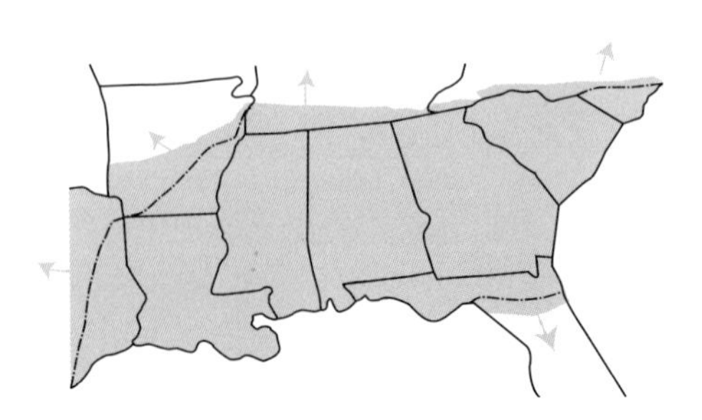

Mesadenus

Mesadenus is a small subtropical and tropical genus with eight species. Like many others of the segregate spiranthoid genera, it is typified by having its leaves in small basal rosettes that wither at or just after flowering. The minute flowers are often presented in one-sided (secund) spikes and the petals and sepals are barely differentiated. We have but a single species in the Southeast, the **copper ladies'-tresses**, *M. lucayanus*, and it occurs in Florida.

Mesadenus lucayanus (Britton) Schlechter

copper ladies'-tresses

Florida; West Indies, Mexico, Central America
Florida: known in our area from a single site
Plant: terrestrial, spike to 40 cm tall, very slender and delicate
Leaves: 2–5; in a basal rosette, 1–3 cm wide × 3–6 cm long, withering at flowering time
Flowers: 8–60; petals and sepals similar, lanceolate, coppery-green; lip rosy green, oblanceolate, channeled or trough-shaped; individual flower size 4–6 mm
Habitat: dry, calcareous woodlands usually with live oak, juniper, and exposed limestone
Flowering period: late January–March

One of the rarest of the *Spiranthes* segregate genera, the copper ladies'-tresses is known from only a few sites in Florida with the northernmost being at Ft. George State Park east of Jacksonville. Of all of these spiranthoid segregates, this slender and delicate species is perhaps the most easily overlooked. The coppery flowers blend in with surrounding leaves on the ground. The plants flower in midwinter or early spring, when not much else is lush and green, with coloring like the rest of the seasonal landscape. It is definitely a "now you see it now you don't" plant. There has been a great deal of confusion as to whether we have *Mesadenus polyanthus* or *M. lucayanus* in Florida, and recent work (Brown 2000) has solved the situation. Plants of *M. polyanthus* are found at high elevations, primarily on volcanic soils, in and around Mexico City, whereas *M. lucayanus* is a lowlands species of scrubby, often calcareous oak woodlands, that is widespread in southern Mexico, Central America, and the West Indies, as well as occurring sparingly in Florida.

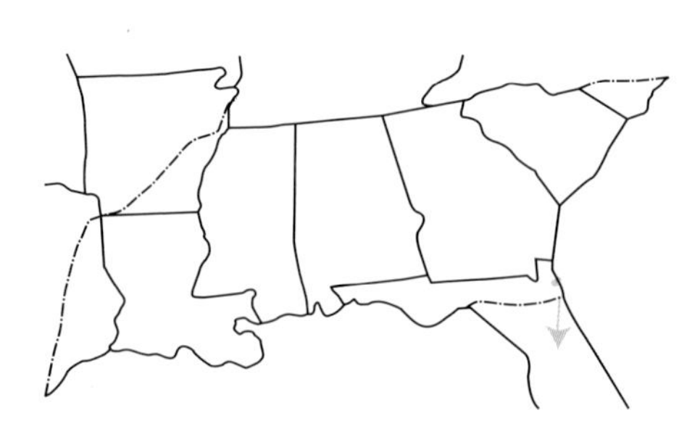

Platanthera

The genus *Platanthera* comprises about 40 North American and Eurasian species, primarily of temperate climes, and is one of the major segregate genera traditionally placed by many botanists within *Habenaria*. It is the largest genus of orchids in the United States and Canada and, with 13 species found in the Southeast, including one variety and several hybrids, the second largest genus in this region. The plants are distinguished from *Habenaria* by their lack of both basal rosettes and tubers or tuberoid roots. Many of the species have large, colorful, showy flowers in tall spikes or racemes. There are several sections to the genus but the most prominent in the Southeast is the section Blephariglottis, the fringed orchises. There are two groups within this section: those species with an entire, or unlobed, lip, and those with a 3-parted lip.

Note: nearly all species of *Platanthera* can be found in both full sun and deeply shaded habitats. Plants in the sun tend to be shorter, to have more densely flowered inflorescences, and to have more upright leaves, whereas those growing in shaded areas tend to be taller and have elongated, loosely flowered inflorescences with spreading leaves. The individual flower size remains the same, but the overall appearance of the plants can be markedly different—to the point that some observers initially think they have two different species!

1a lip margin entire; flowers greenish in color . . . 12
1b lip margin variously fringed, eroded, lacerated, or toothed . . . 2

2a lip 3-lobed . . . 8
2b lip unlobed (entire), the margin variously fringed, eroded, lacerated or toothed . . . 3

3a lip conspicuously fringed . . . 4
3b lip erose or lacerated but not fringed; flowers white . . . **monkey-face orchis,** *P. integrilabia*

4a flowers white . . . 5
4b flowers yellow or orange . . . 6

5a lip descending to recurved, with a very short isthmus; fringe lacerate; spur usually less than 26 mm . . . **northern white fringed orchis,** *P. blephariglottis*
5b lip projecting, narrowed to an extended isthmus; fringe delicately filiform; spur usually exceeding 30 mm; coastal plain . . . **southern white fringed orchis,** *P. conspicua*

6a spur greatly exceeding the lip; 2.5–3.5 cm long; column tapering to a point . . . **orange fringed orchis,** *P. ciliaris*

6b spur equal to or less than the lip . . . 7

7a spur 0.4–1.0 cm long; spur orifice circular; column face flattened . . . **orange crested orchis,** *P. cristata*

7b spur 0.8–1.4 cm long; spur orifice keyhole-shaped; column bent forward with a pronounced hook . . . **Chapman's fringed orchis,** *P. chapmanii*

8a petals entire, lip deeply lacerate; flowers greenish white to creamy green . . . **green fringed orchis,** *P. lacera*

8b petals erose or shallowly fringed at the apex . . . 9

9a lip merely erose . . . **purple fringeless orchis,** *P. peramoena*

9b lip distinctly fringed or lacerate . . . 10

10a flowers typically in shades of purple . . . 11

10b flowers white . . . **eastern prairie fringed orchis,** *P. leucophaea*

11a lip margin fringed to more than 1/3 the length of the lip; spur orifice circular . . . **large purple fringed orchis,** *P. grandiflora*

11b lip margin fringed less than 1/3 the length of the lip, spur orifice a transverse dumbbell or bowtie-shaped . . . **small purple fringed orchis,** *P. psycodes*

12a lip oval; floral bracts usually equal to or shorter than the flowers; plants primarily of the southeastern states . . . **southern tubercled orchis,** *P. flava* var. *flava*

12b lip oblong; floral bracts longer than the flowers; plants primarily of the central and northeastern states . . . **northern tubercled orchis,** *P. flava* var. *herbiola*

Platanthera blephariglottis (Willdenow) Lindley

northern white fringed orchis

Michigan east to Newfoundland, and south to Georgia

forma *holopetala* (Lindley) P. M. Brown—entire-lip form

Genera and Species of Orchidaceous Plants 291. 1835, as *Platanthera holopetala*, type: Canada.

Georgia, North Carolina, South Carolina: restricted to a few locations on the northern edge of the Coastal Plain or lower Piedmont
Plant: terrestrial, 25–60 cm tall
Leaves: 2–4; cauline, lanceolate 2–4 cm wide × 8–20 cm long, gradually reduced to bracts within the inflorescence
Flowers: 20–45; arranged in a dense terminal raceme; sepals ovate, petals linear, enclosed within the sepals forming a hood; lip spatulate with a coarse, fringed margin or, in the forma *holopetala*, the margin slightly if at all fringed; lip narrowed to an obscure isthmus at the base; perianth pure white; individual flower size 1.75–2.0 cm, not including the 1.5–2.0 cm spur
Habitat: open wet meadows, roadside ditches and seeps, and pine flatwoods
Flowering period: August–October

This large, showy member of the fringed-lipped section Blephariglottis is widespread throughout eastern North America and reaches the southern extreme of its distribution in a few inland counties of the southeastern United States. It, like several other species of *Platanthera*, is affected by habitat succession and often flowers best a few years after the disturbance of adjacent woody plant material—i.e., burning, hurricanes, mowing, etc.

There has been much confusion over the years pertaining to this species and its sister species *Platanthera conspicua*. *Platanthera blephariglottis* is essentially a northern species that barely dips southward into the southeastern United States. Careful comparison of the lip criteria will always aid in identifying these two species. Range can also be helpful as the overlap is minimal and the flowering times tend to be somewhat different, with *P. blephariglottis* flowering earlier than *P. conspicua*.

The most important thing to remember is that there are large-flowered plants of *P. blephariglottis* in the central Atlantic states and they have often been mistaken for *P. conspicua* primarily because of earlier literature that simply gave the range for the large-flowered plants, identified as *P. blephariglottis* var. *conspicua*, as north to Cape Cod along the Coastal Plain. Should you encounter plants in that small area of overlap look very carefully at the shape of the lip and the fringe as well as the positioning of the flowers.

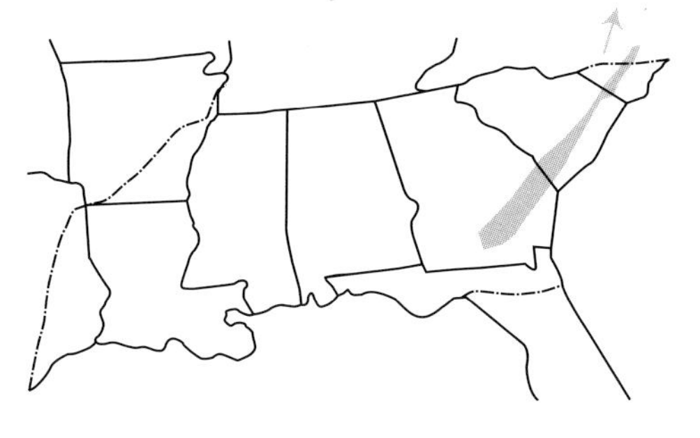

Platanthera chapmanii (Small) Luer *emend.* Folsom

Chapman's fringed orchis

southeastern Georgia, Florida, and eastern Texas
Texas, Georgia, Florida: a local endemic that is frequent only in northern Florida
Globally Threatened
Plant: terrestrial, to 100(110) cm tall
Leaves: 2–4; cauline, lanceolate 0.75–4.0 cm wide × 5–25 cm long, rapidly reduced to bracts within the inflorescence
Flowers: 30–75(92); arranged in a dense terminal raceme; sepals ovate, petals linear, the apex erose to slightly fringed, enclosed within the sepals forming a hood; lip ovate with a delicately fringed margin; perianth brilliant orange; individual flower size 2 cm, not including the 0.8–1.7 cm spur that typically is parallel to the ovary; column bent forward with a pronounced hook; orifice keyhole-shaped
Habitat: open wet meadows, roadside ditches and seeps, and pine/palmetto flatwoods
Flowering period: late July–early September

Not always easy to identify, Chapman's fringed orchis ancestrally arose as a hybrid between *Platanthera ciliaris*, the **orange fringed orchis**, and *P. cristata*, the **orange crested orchis.** Many years of adaptation have resulted in a pollinator-specific, stable, reproducing species. Many texts still cite it as *P.* ×*chapmanii*, referring to its perceived hybrid status (see Folsom, 1984; Brown, 2003 for details). Because of its origins, this species falls morphologically between the two ancestors. The two best characters are the spur that is equal in length to the lip, and the bent column. Also, unlike the true contemporary hybrid of *P. ciliaris* and *P. cristata*—*P.* ×*channellii*, *P. chapmanii* occurs in pure stands. From a global standpoint this is one of the rarest orchids we know. It has recently been classified as G1G2 by The Nature Conservancy. *P. chapmanii* is known from a single (historical) site in southern Georgia and several scattered, although substantial, sites in northern Florida, and then jumps to a few small stands in four counties of eastern Texas, where formerly it was more abundant. More than 90 percent of the known plants in the world are found within the national forests of northern Florida. Hybrids with *P. cristata* are known as *P.* ×*apalachicola* and those with *P. ciliaris* as *P.* ×*osceola*. See pages 168–170 for additional information.

Platanthera ciliaris (Linnaeus) Lindley

orange fringed orchis

southern Michigan east to Massachusetts, and south to Florida and Texas
Texas, Louisiana, Arkansas, Alabama, Mississippi, Georgia, Florida, South Carolina, North Carolina: locally common throughout
Plant: terrestrial, 25–100+ cm tall
Leaves: 2–5; cauline, lanceolate 1–5 cm wide × 5–30 cm long, gradually reduced to bracts within the inflorescence
Flowers: 30–75; arranged in a dense terminal raceme; sepals ovate, petals linear, fringed at the tip, enclosed within the sepals forming a hood; lip ovate with a delicately fringed, filiform margin; perianth deep yellow to orange; individual flower size 4 cm, not including the 2.5–3.5 cm spur that typically is loosely descending; the column tapering to a point
Habitat: open wet meadows, roadside ditches and seeps, and pine flatwoods
Flowering period: late July–late September

A near mirror image of the **southern white fringed orchis,** *Platanthera conspicua*, the brilliant, deep yellow to orange plumes of the orange fringed orchis can be a meter in height. Scattered populations are well known from all of the states in the Southeast. With a preference for areas that stay wet to damp in the hot, dry days of summer, populations vary from year to year in how thrifty they are. This species often grows with both of its white cousins and the hybrids *P.* ×*bicolor* and *P.* ×*lueri* can often can be found as well. Within populations in both the Apalachicola and Osceola National Forests hybrid swarms can also be seen that involve *P. cristata* and *P. chapmanii* and often result in numerous backcrosses. Hybrids with *P. cristata* are known as *P.* ×*channellii* and those with *P. chapmanii* as *P.* ×*osceola*. See pages 168–70 for additional information.

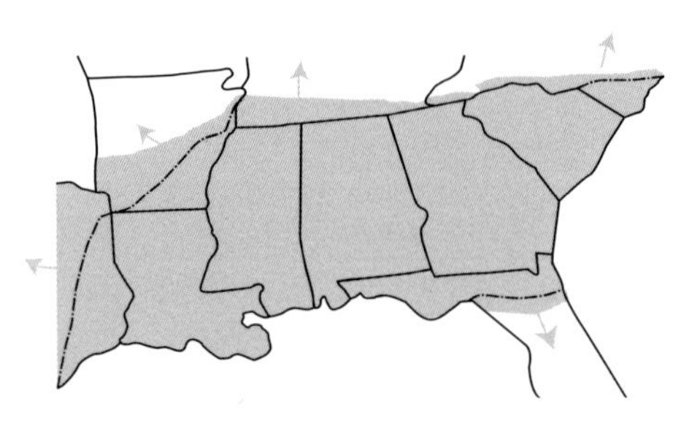

Platanthera conspicua (Nash) P. M. Brown

southern white fringed orchis

southeastern North Carolina, south along the Coastal Plain to eastern Texas
Texas, Louisiana, Alabama, Mississippi, Georgia, Florida, South Carolina, North Carolina: widespread and locally common on the Atlantic Coastal Plain; rare and local westward; extirpated in Texas
Plant: terrestrial, 25–100+ cm tall
Leaves: 2–4 cauline, lanceolate 1–5 cm wide × 5–35 cm long, rapidly reduced to bracts within the inflorescence
Flowers: 30–65; arranged in a dense terminal raceme; sepals ovate, petals linear, enclosed within the sepals forming a hood; lip ovate with a delicately fringed margin narrowed to a distinct isthmus at the base; perianth pure white; individual flower size 3 cm, not including the 3–4 cm spur
Habitat: open wet meadows, roadside ditches and seeps, and pine flatwoods
Flowering period: August–October

The fringed orchids of eastern and central North America present some of the showiest orchids of the summer and the stately, snow-white plumes of the southern white fringed orchis are no exception. Widely scattered throughout the southeastern Coastal Plain, this species, like other species of fringed orchids, has fallen victim to construction. Although it prefers open damp meadows, pine flatwoods, and seeps, it is now most frequently found in narrow roadside ditches and open sphagnous areas within the woodlands. Plants flower over a long period of time in mid– to late summer. See the entry for *Platanthera* hybrids on pages 164–66 for those individuals that are lemon or pale cinnamon colored.

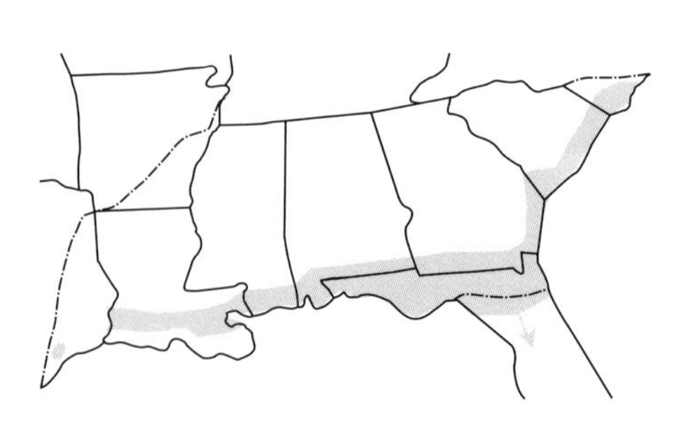

Platanthera cristata (Michaux) Lindley

orange crested orchis

southeastern Massachusetts south to Florida, and west to eastern Texas; primarily on the Coastal Plain

forma *straminea* P. M. Brown—pale yellow-flowered form
North American Native Orchid Journal 1(1): 12. 1995, type: New Jersey

Texas, Louisiana, Arkansas, Alabama, Mississippi, Georgia, Florida, South Carolina, North Carolina: widespread and scattered throughout our area, primarily on the Coastal Plain region
Plant: terrestrial, to 80 cm tall
Leaves: 2–4; cauline, lanceolate, 1–3 cm wide × 5–20 cm long, rapidly reduced to bracts within the inflorescence
Flowers: 30–80; arranged in a loose to dense terminal raceme; sepals ovate, petals spatulate with margin of the apex crested, enclosed within the sepals forming a hood; lip long triangular-ovate with a coarsely lacerate margin; perianth deep yellow to orange or, in the forma *straminea*, pale yellow; individual flower size 5–7 mm, not including the 7 mm spur that is typically curved; the column is very short and the face flattened
Habitat: open wet meadows, roadside ditches and seeps, and pine flatwoods
Flowering period: late June–late September

The orange crested orchis is a smaller, and perhaps more refined version of the **orange fringed orchis,** *Platanthera ciliaris.* It is more widespread throughout the Coastal Plain and often occurs without other related species nearby. Open pine flatwoods is its preferred habitat, although it can also be found in damp meadows and flatwoods, ditches, and roadside seeps. The raceme is usually about 2.5 cm in diameter and the spur is always shorter than the lip and typically curved forward, whereas in *P. ciliaris* the loosely descending spur is much longer than the lip and the raceme is 4.5+ cm in diameter.

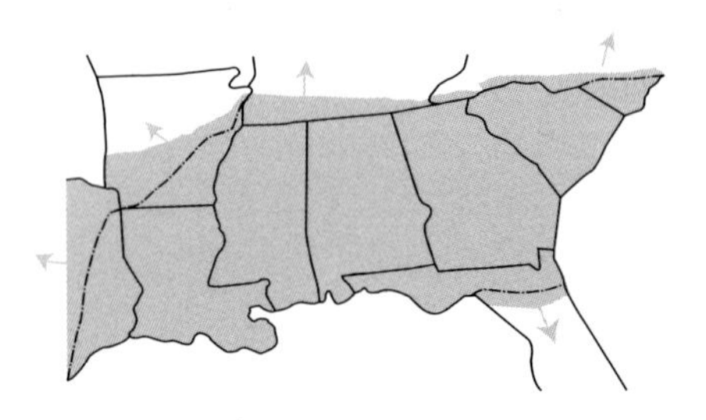

forma *straminea*

Platanthera flava (Linnaeus) Lindley var. *flava*

southern tubercled orchis

southwestern Nova Scotia; Missouri east to Maryland, south to Florida and Texas
Texas, Louisiana, Arkansas, Alabama, Mississippi, Georgia, Florida, South Carolina, North Carolina: local and scattered throughout the area; extending well north and occasionally overlapping with the var. *herbiola*
Plant: terrestrial, 10–60 cm tall
Leaves: 2–4; cauline, nearly basal, lanceolate, 1–4 cm wide × 5–20 cm long, rapidly reduced to bracts within the inflorescence
Flowers: 10–40; arranged in a loose to dense terminal raceme; sepals and petals ovate, enclosed within the dorsal sepal forming a hood; lip ovate with a prominent tubercle in the center; perianth yellow-green; individual flower size 6–7 mm, not including the 8 mm spur
Habitat: open wet meadows, roadside ditches and seeps, swamps and shaded floodplains
Flowering period: late April–July

This species is equally at home in shaded, wet woodlands, streamsides, and bright sunny, open, damp roadsides. Although the flowers are identical in both habitats, the habit of the plant varies greatly. Those in shaded habitats tend to be tall and slender, with flowers spaced out along the stem, whereas those in sunnier habitats have flower spikes and leaves that are very compact and crowded. The flowers in the shade tend to be very green in color and those in the sun much more yellow, tending toward chartreuse. But in both instances that distinctive tubercled lip is always prominent.

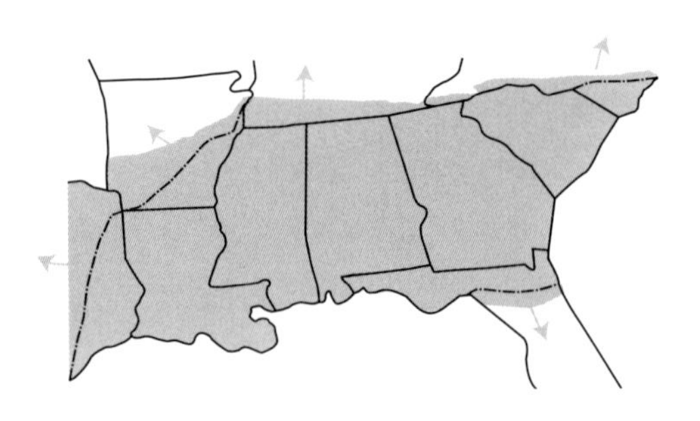

Platanthera flava (Linnaeus) Lindley var. *herbiola* (R. Brown) Luer

northern tubercled orchis

Minnesota east to Nova Scotia, south to Missouri and Georgia; south in the Appalachian Mountains

forma *lutea* (Boivin) Whiting & Catling—yellow-flowered form

Naturaliste Canadien. 94(1): 146. 1967, as *Habenaria flava* var. *herbiola* forma *lutea*, type: Ontario

Georgia: rare in northern counties; a recent unvouchered report from South Carolina
Plant: terrestrial, 10–50 cm tall
Leaves: 3–5; cauline, nearly basal, lanceolate 2–5 cm wide × 8–20 cm long, rapidly reduced to bracts within the inflorescence
Flowers: 15–45; arranged in a loose to dense terminal raceme; sepals and petals ovate, enclosed within the dorsal sepal forming a hood; lip oblong with a prominent tubercle near the base and triangular lobes on either side; perianth grassy-green or, in the forma *lutea*, distinctly yellow; individual flower size 6–7 mm, not including the 5–7 mm spur
Habitat: open wet meadows, roadside ditches and seeps, swamps and shaded floodplains
Flowering period: late May–June

Platanthera flava var. *herbiola*, an obscure, grass-green, fragrant species, was once considered to be one of the rarer orchids in eastern North America. As the result of intensive field searches by many individuals we now know of some large, stable populations in several states. The northern variety of the tubercled orchis is at the southeastern extreme of its range in northern Georgia. A recent unverified report for adjacent South Carolina is consistent with that range. Its sweet, perfume-like fragrance is often detected before the plant is actually seen. The nominate variety *flava* occurs primarily in the southeastern and south-central United States with disjunct populations in southwestern Nova Scotia and southern Ontario.

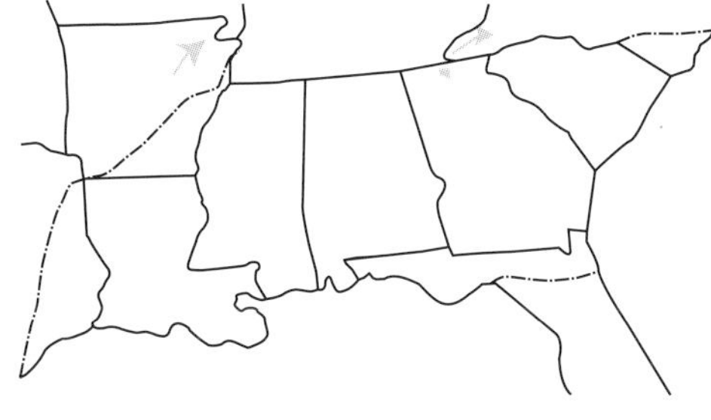

Platanthera grandiflora (Bigelow) Lindley

large purple fringed orchis

Ontario east to Newfoundland, south to West Virginia and New Jersey; south in the Appalachian Mountains to northern Georgia

forma *albiflora* (Rand & Redfield) Catling—white-flowered form

Flora of Mt. Desert Island 152. 1894, as *Habenaria fimbriata* forma *albiflora*, type: Maine

forma *bicolor* P. M. Brown—bicolor-flowered form

North American Native Orchid Journal 1(4): 289. 1995, type: Newfoundland

forma *carnea* P. M. Brown—pink-flowered form

North American Native Orchid Journal 1(4): 289. 1995, type: Newfoundland

forma *mentotonsa* (Fernald) P. M. Brown—entire-lip form

Rhodora 48: 184. 1946, type: Nova Scotia

Georgia: very rare at the southern limit of its range
Plant: terrestrial, to 120 cm tall
Leaves: 2–6; cauline, lanceolate, keeled 1.5–7.0 cm wide × 8–24 cm long, gradually reduced to bracts within the inflorescence
Flowers: 30–65; arranged in a loose to dense terminal raceme usually 3–5 cm in diameter with eventually all of the flowers open simultaneously; sepals ovate, petals spatulate with dentate margins; lip 3-parted with a coarsely fringed margin usually to more than 1/3 the depth of the lip or, in the forma *mentotonsa*, the margin essentially entire; perianth various shades of purple from pale lavender to deep, rich magenta or, in the forma *albiflora*, white or, in the forma *bicolor*, purple and white or, in the forma *carnea*, a delicate fresh pink; individual flower size 2–3 cm, not including the 2.5 cm spur; spur orifice rounded
Habitat: open wet meadows, roadside ditches and seeps, mountain glades
Flowering period: late June–early July

The large purple fringed orchis is widespread throughout much of central and northeastern North America. This tall (to 1 meter), stately species is often a feature of the early summer landscape. It usually occurs in small numbers, often as an individual plant, but every once in a while large stands of more than 100 plants can be found. Because *Platanthera grandiflora* is reaching the southern extreme of its range in Gilmer County, Georgia, variation may be minimal. Although similar in overall appearance to the **small purple fringed orchis,** *P. psycodes,* this species has several points that will aid in identification. Be sure to carefully note the shape of the orifice, depth of the fringing, and overall shape of the inflorescence. Hybrids with *P. lacera* are known as *P. ×keenanii.*

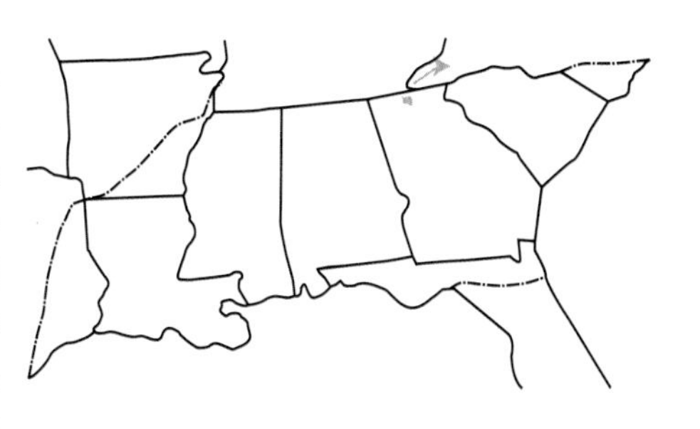

forma *albiflora*

with *P. ×keenanii,* white form

Platanthera integrilabia (Correll) Luer

monkey-face orchis

Kentucky and North Carolina, south to Mississippi and Georgia
Alabama, Mississippi, Georgia, South Carolina: known only from a few, very restricted sites; may be extirpated in South Carolina
Globally Threatened
Plant: terrestrial, 25–60 cm tall
Leaves: 2–3; cauline, lanceolate, keeled, 2–3 cm wide × 8–20 cm long, rapidly reduced to bracts within the inflorescence
Flowers: 6–15; arranged in a loose terminal raceme; sepals ovate, petals oblong, the margins entire, nearly enclosed within the dorsal sepal forming a hood; lip entire, spatulate-lanceolate, narrowed to a prominent isthmus at the base and the margin finely serrate; perianth pure white; individual flower size ca. 3 cm, not including the 4–5 cm strongly curved spur
Habitat: shaded, damp slopes and ravines
Flowering period: August

The monkey-face orchis is one of the most geographically restricted species we have in North America. Endemic to the southern Appalachians and adjacent Cumberland Plateau, most sites are very small and imperiled. The only site of any size is just beyond the range of this work in southeastern Tennessee at Starr Mountain. Plants of this species have been known for many years, but it was not until 1941 that Donovan Correll described the species. *Platanthera integrilabia* was first collected from the Cumberland Mountains of Tennessee in 1888, and was noted by Ames in 1910 as a form of *Habenaria blephariglottis* var. *conspicua* with an entire labellum. *P. integrilabia* produces large quantities of single-leaved seedlings compared to the number of mature flowering plants. Although plants are often found growing with *P. cristata*, *P. ciliaris*, and *Gymnadeniopsis clavellata*, no intergradations and/or hybrids have ever been noted.

Platanthera lacera (Michaux) G. Don

green fringed orchis, ragged orchis

Manitoba east to Newfoundland, south to Texas and Georgia
Texas, Arkansas, Alabama, Mississippi, Georgia, South Carolina, North Carolina: rare and local in the Southeast, becoming more frequent northward
Plant: terrestrial, 20–80 cm tall
Leaves: 3–6; cauline, lanceolate, keeled, 2.5–5.0 cm wide × 8–24 cm long, gradually reduced to bracts within the inflorescence
Flowers: 12–40, highly variable; arranged in a loose to dense terminal raceme; sepals obovate, the petals oblong, upright, usually with entire margins; lip 3-parted and deeply lacerate; perianth various shades of green; individual flower size ca. 1.5–3.0 cm, not including the 16–23 mm spur, the orifice nearly square
Habitat: open wet meadows, roadside ditches and seeps, mountain meadows
Flowering period: late June–early July

The least conspicuous of the fringed orchises, *Platanthera lacera* is scattered throughout the Piedmont and mountains of the Southeast. It can be found during the early summer months in damp meadows, open wet woods, and roadside ditches. Flower color is highly variable in many shades of green and some plants are nearly white. Those plants whose flowers show a wash of lavender may represent hybrids with either *P. psycodes* (*P.* ×*andrewsii*) or *P. grandiflora* (*P.* ×*keenanii*). Because of the great rarity of either of the purple-flowered parents in the Southeast the possibility of the hybrids is lessened.

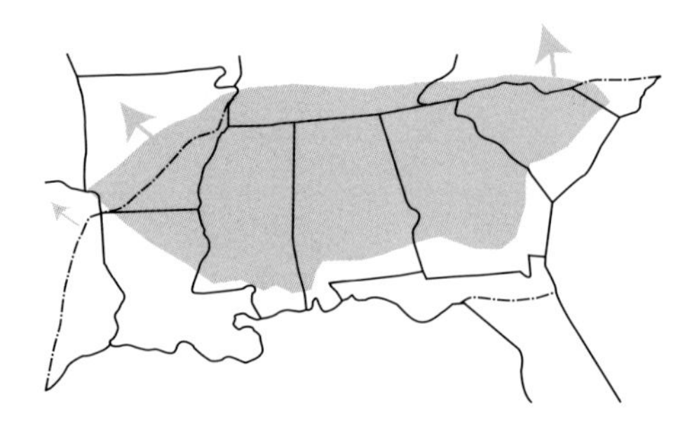

Platanthera leucophaea (Nuttall) Lindley

eastern prairie fringed orchis

Nebraska east to Ontario and Maine, south to Oklahoma, Louisiana, and Virginia
FEDERALLY LISTED AS THREATENED
Louisiana: known only from an old herbarium record; presumed extirpated
Plant: terrestrial, 50–120 cm tall
Leaves: 2–5; cauline, lanceolate, keeled, 2.5–5.0 cm wide × 8–20 cm long, rapidly reduced to bracts within the inflorescence
Flowers: 12–27; arranged in a loose to dense terminal raceme; sepals ovate, the petals spatulate and broadly rounded, the margin finely serrate, partially enclosed with the sepals to form a cup-shaped hood; lip 3-parted and deeply fringed; perianth cream-colored, shading to whitish-green; individual flower size ca. 2–3 cm, not including the 3–5 cm spur, the orifice broadly crescent-shaped
Habitat: wet prairies, fens
Flowering period: June?

The inclusion of the eastern prairie fringed orchis in the orchid flora of the southeastern United States is based upon an old specimen labeled Louisiana. No further details are on the label. Sheviak (1987) gives more corroborating details that support this record. Living plants have not been seen here in recent history and the nearest extant population is a disjunct site several hundred miles away in Oklahoma. Little habitat still exists in Louisiana for this species. The plants are unmistakable and could be confused only with the **western prairie fringed orchis,** *Platanthera praeclara*, which has not been found in the Southeast. The large, showy flowers are exceedingly fragrant and pollinated by sphinx moths.

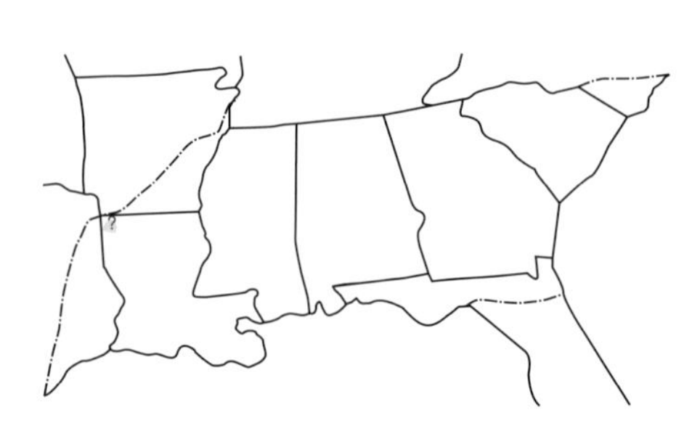

Platanthera peramoena (A. Gray) A. Gray

purple fringeless orchis

Missouri east to New Jersey, south to Mississippi and Georgia
forma *doddiae* P. M. Brown
North American Native Orchid Journal 8: 30–31. 2002, type: Missouri

Arkansas, Alabama, Mississippi, Georgia, South Carolina: rare and local in the northern counties, more frequent in central Mississippi
Plant: terrestrial, 75–105 cm tall
Leaves: 2–5; cauline, elliptic-lanceolate, keeled, 2.5–5.0 cm wide × 8–14 cm long, gradually reduced to bracts within the inflorescence
Flowers: 27–50+; arranged in a dense terminal raceme; sepals elliptic, reflexed; petals spatulate and rounded, the margin finely serrate, partially enclosed with the dorsal sepal to form a hood; lip 3-parted, apically notched and faintly erose; perianth a brilliant rosy-purple or, in the forma *doddiae*, pure white; individual flower size ca. 2–3 cm, not including the 2.5–3.0 cm spur, the orifice rounded
Habitat: damp meadows, low wet woods, streamsides
Flowering period: midsummer; late June–early August

One of the tallest and most striking of the fringed orchises, *Platanthera peramoena* is a characteristic species of the east-central United States. Not necessarily common anywhere, it can be locally abundant. Here in the Southeast it barely reaches into the northern border counties of South Carolina, Georgia, and Alabama, and is more widespread into central Mississippi and northeastern Arkansas. Encountering these plants in flower is a real delight; despite careful searching, the white-flowered form has been reported only once, and then from Missouri. Not easy to overlook, it should be carefully sought elsewhere.

Platanthera psycodes (Linnaeus) Lindley

small purple fringed orchis

Ontario east to Newfoundland, south to West Virginia and New Jersey; south in the Appalachian Mountains to Georgia

forma *albiflora* (R. Hoffman) Whiting & Catling—white-flowered form

Proceedings of the Boston Society of Natural History 36: 248. 1922, as *Habenaria psycodes* forma *albiflora*, type: Massachusetts

forma *ecalcarata* (Bryan) P. M. Brown—spurless form

Annals of the Missouri Botanical Garden 4: 38. 1917, as *Habenaria psycodes* var. *ecalcarata*, type: Michigan

forma *rosea* P. M. Brown—pink-flowered form

North American Native Orchid Journal 1(4): 289. 1995, type: Vermont

forma *varians* (Bryan) P. M. Brown—entire-lip form

Annals of the Missouri Botanical Garden 4: 37. 1917, as *Habenaria psycodes* var. *varians*, type: Michigan

Georgia, South Carolina: rare and local in two northern counties
Plant: terrestrial, to 90 cm tall
Leaves: 2–6; cauline, lanceolate, keeled 1.5–7.0 cm wide × 8–24 cm long, gradually reduced to bracts within the inflorescence
Flowers: 30–125; arranged in a loose to dense terminal raceme usually 2.5–3.0 cm in diameter with flowers opening successively, i.e. the lower one usually withering before the upper ones have opened, giving the inflorescence a conical appearance; sepals elliptic, petals obovate with finely dentate margins; lip 3-parted with a finely fringed margin usually to less than 1/3 the depth of the lip, or in the forma *varians* the margin essentially entire; perianth various shades of purple from pale lavender to deep, rich rosy-magenta or, in the forma *albiflora*, white or, in the forma *rosea*, a pale pink; individual flower size 0.5–1.5 cm, not including the 1.2–1.8 cm spur, or in the forma *ecalcarata*, the spur lacking; spur orifice likened to a transverse dumbbell
Habitat: open wet meadows, roadside ditches and seeps, mountain meadows
Flowering period: late June–early July

The common names large and small purple fringed orchis are very misleading: the small purple fringed orchis can often be larger than the large purple fringed orchis. The small purple fringed orchis is usually both taller and more floriferous than *Platanthera grandiflora*, although the individual flowers are smaller. The small purple fringed orchis is also widespread throughout much of central and northeastern North America. This tall, slender species is at home in open meadows as well as wooded streamsides. It usually occurs in small numbers, but is rarely found as a single plant. Because it is reaching the southern extreme of its range in the mountains of northern Georgia and South Carolina, variation may be minimal. For comparisons to the **large purple fringed orchis,** *P. grandiflora*, see details at that entry on page 150. Hybrids with *P. lacera* are known as *P.* ×*andrewsii*.

forma *albiflora*

Hybrids:

Platanthera ×andrewsii (Niles) Luer

Andrews' hybrid fringed orchis

(*P. lacera × P. psycodes*)

Bog-trotting for Orchids. 258, 1904, as *Habenaria ×andrewsii*, type: Vermont

Andrews' hybrid fringed orchis is not uncommon throughout most of the range of the two parents, but because of the rarity of *Platanthera psycodes* in the Southeast the hybrid is less likely.

Platanthera ×apalachicola P. M. Brown & S. Stewart

Apalachicola hybrid fringed orchis

(*P. chapmanii × P. cristata*)

North American Native Orchid Journal 9: 35. 2003, type: Florida

Platanthera ×apalachicola is locally common in northern Florida where both parents frequently grow together. The hybrids usually occur as individuals and may appear within stands of *P. chapmanii* as smaller flowered, more slender plants or within stands of *P. cristata* as larger flowered, more robust individuals. The hooked column of *P. chapmanii* is usually dominant but the spur length and position are intermediate.

Platanthera ×beckneri P. M. Brown
Beckner's hybrid fringed orchis
(*P. conspicua* × *P. cristata*)
North American Native Orchid Journal 8: 3–14. 2002, type: Florida

These small, pale yellow to cream-colored plants of *Platanthera ×beckneri* are often found in hybrid swarms.

Platanthera ×bicolor (Rafinesque) Luer
bicolor hybrid fringed orchis
(*P. blephariglottis* × *P. ciliaris*)
Flora Telleuriana. 2: 39. 1839, as *Blephariglottis bicolor,* type: New Jersey

Plants of the bicolor hybrid fringed orchis usually have lemon, pale tan, or bicolored flowers and retain the overall dimensions of the parents. The shape of the lip, lacking the elongated isthmus of *Platanthera conspicua*, will help to confirm this taxon.

Platanthera ×*canbyi* (Ames) Luer
Canby's hybrid fringed orchis
(*P. blephariglottis* × *P. cristata*)
Rhodora 10: 70. 1908, as *Habenaria canbyi*, type: Delaware

Plants of *Platanthera* ×*canbyi* vary in color from nearly white to pale yellow to light orange. The spur is usually about as long as the lip and the lack of the elongated isthmus of *P. conspicua* will confirm identification as compared to those of *P.* ×*beckneri*.

Platanthera ×*channellii* Folsom
Channell's hybrid fringed orchis
(*P. ciliaris* × *P. cristata*)
Orquidea (Mex.) 9(2): 344. 1984, type: Alabama

This hybrid and the species *Platanthera chapmanii* can be difficult to tell apart. One of the best ways is to look about and see which other species are growing nearby. If all the plants observed are the same, and within the range of *P. chapmanii*, it is most likely *P. chapmanii*, whereas if it is a colony of mixed species and only a few intermediate plants are present, it is more likely *P.* ×*channellii*.

Platanthera ×*keenanii* P. M. Brown
Keenan's hybrid fringed orchis
(*P. grandiflora* × *P. lacera*)
A Field & Study Guide to the Orchids of New England and New York. 189. 1994, type: Maine

Keenan's hybrid fringed orchis is even less likely than *Platanthera* ×*andrewsii* in the Southeast because of the extreme rarity of *P. grandiflora.*

Platanthera ×*lueri* P. M. Brown
Luer's hybrid fringed orchis
(*P. conspicua* × *P. ciliaris*)
North American Native Orchid Journal 8: 3–14. 2002, type: Florida

Platanthera ×*lueri* is the Southeastern counterpart to the more northern and inland *Platanthera* ×*bicolor* with a similar range of color, from rich cream and lemon to coffee. Again watch for that narrowed isthmus indicating the *P. conspicua* parent.

Platanthera ×*osceola* P. M. Brown & S. Stewart
Osceola hybrid fringed orchis
(*P. chapmanii* × *P. ciliaris*)
North American Native Orchid Journal 9: 35. 2003, type: Florida

Platanthera ×*osceola* is known only from the Osceola National Forest, where in a very few local sites it is the only documented area where both parents are found growing together. Plants of the hybrid usually occur as individuals and may appear within stands of *Platanthera chapmanii* as larger flowered, more robust plants with decidedly longer spurs or within stands of *P. ciliaris* as smaller more compactly flowered individuals. The hooked column of *P. chapmanii* is not as dominant as in *P.* ×*apalachicola*.

×*Platanthopsis vossii* (Case) P. M. Brown
Voss' hybrid rein orchis
(*Gymnadeniopsis clavellata* × *Platanthera blephariglottis*)
Michigan Botanist. 22: 141–144.1983, as *Platanthera* ×*vossii,* type: Michigan.

See page 92 for a full account of this rare intergeneric hybrid.

Understanding *Platanthera chapmanii*, Its Origins and Hybrids

Note: some of the information below is repeated from other places within this publication. Because of the confusion in the identification of *Platanthera chapmanii*, this repetition is intentional.

Although restricted to the southern portion of the southeastern United States, *Platanthera chapmanii*, Chapman's fringed orchis, is an important component of the summer-flowering orchid flora of the Gulf Coastal Plain and northeastern Florida. Historically known from eastern Texas, much of northern Florida and a single site in southeastern Georgia, today it can be best found in the Apalachicola and Osceola National Forests of Florida. Other small sites in northern Florida persist. The species is absent from the eastern half of the Panhandle, and the Marion and Polk County records for Florida appear to be *P.* ×*channellii*. Only a few sites remain in eastern Texas and the Georgia locale is based upon a historic collection. No collections have ever been made from the area between Apalachicola and East Texas.

Understanding this species and its relationships to *Platanthera ciliaris* and *P. cristata* is greatly simplified if the observer can see all three taxa in one field session. This can be accomplished only in the Osceola National Forest, for *P. ciliaris* is historically and apparently absent from any of the other known localities for *P. chapmanii*. Liggio and Liggio (1999) clearly state that *P. ciliaris* has never been found within any of the Texas locales for *P. chapmanii*. Conversely, *P. cristata* is often found growing within or near many of the *P. chapmanii* sites, especially in eastern Florida.

Folsom aptly demonstrated in his 1984 publication that the origins of *Platanthera chapmanii* were most likely an ancient hybridization of *P. ciliaris* and *P. cristata*. Therefore *P. chapmanii* appears to be intermediate in size and characters between the two ancestors. Over the years it has evolved into a stable, reproducing species with a very distinctive column. At the same time, the contemporary hybrid of *P. ciliaris* and *P. cristata*, *P.* ×*channellii*, occurs in rare situations when both parents are present. It, too, is intermediate between the parents, but the column is unlike that of *P. chapmanii*. One of the best helps in the initial determination of plants in the field is observing what predominates in the area. If both *P. ciliaris* and *P. cristata* are present and only a few intermediates are to be found, then they, in all probability, will be the hybrid, *P.* ×*channellii*. If the majority of plants appear intermediate between *P. ciliaris* and *P. cristata* and only a few of either of the latter species are present, then the observer needs to look carefully at the shape of the column; most likely the majority of plants will be *P. chapmanii*.

Characters that help in determining which species are present include geographic location, diameter of raceme, size of flower, length and position of spur, and shape of orifice. To state simply that *Platanthera ciliaris* is larger, *P. chapmanii*, intermediate in size, and *P. cristata*, smaller, has led to much confusion. For many orchid enthusiasts this reference to size implies overall size, especially height, although height is not mentioned. This is not accurate; height should

P. cristata *P. chapmanii* *P. ciliaris*

P. chapmanii

P. cristata *P. ciliaris*

never be taken into account. All three species can grow from 10 or 15 cm to, in the case of *P. chapmanii* and *P. ciliaris*, more than a meter in height! When size comparisons are made they refer to the diameter of the raceme and measurements of the individual flowers. Even the overall height of the flowering raceme is not a good criterion for identification. Because of the ancestral parentage of *P. chapmanii*, plants can easily favor the overall raceme shape of either parent, but the raceme diameter appears to remain constant. The following illustrations will assist in understanding this comparison.

In addition to understanding the species, orchid observers need to be aware of the hybrids that are readily involved in this complex, including:

Platanthera ×*apalachicola*	*Platanthera* ×*osceola*	*Platanthera* ×*channellii*
(*P. chapmanii* × *P. cristata*)	(*P. chapmanii* × *P. ciliaris*)	(*P. ciliaris* × *P. cristata*)

Relationships among this group are best summed up in the following diagram. The white-flowered species, *Platanthera blephariglottis*, *P. conspicua*, and *P. integrilabia*, are included in this diagram for completeness in the group.

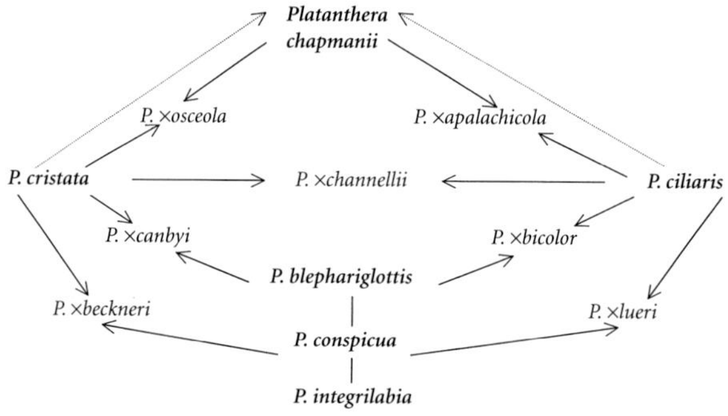

Platanthera ×*apalachicola* is locally common in northern Florida where both parents frequently grow together. The hybrids usually occur as individuals and may appear within stands of *P. chapmanii* as smaller flowered, more slender plants or within stands of *P. cristata* as larger flowered more robust individuals. The hooked column of *P. chapmanii* is usually dominant but the spur length and position are intermediate.

Platanthera ×*channellii* and *P. chapmanii* can be difficult to tell apart. One of the best ways is to look about and see what other species are growing nearby. If all the plants observed are the same, and within the range of *P. chapmanii*, it is most likely *P. chapmanii*, whereas if it is a colony of mixed species and only a few intermediate plants are present it is more likely to be *P.* ×*channellii*.

Platanthera ×*osceola* is known only from Osceola National Forest, the only place documented where both parents are found growing together. Plants of the hybrid usually occur as individuals and may appear within stands of *P. chapmanii* as larger flowered, more robust plants with decidedly longer spurs or within stands of *P. ciliaris* as smaller more compactly flowered individuals. The hooked column of *P. chapmanii* is not as dominant as in *P.* ×*apalachicola*.

Platythelys

Platythelys is a neotropical genus with eight species found primarily in damp woodlands throughout its range. To those who are familiar with the native orchids of temperate North America, *Platythelys* appears as if intermediate between the rattlesnake orchids, *Goodyera*, and the ladies'-tresses, *Spiranthes*, both of which are closely related to *Platythelys*. Several issues have recently been raised concerning the accurate distribution of some species of *Platythelys*. Resolving this will require a re-examination of specimens from the various countries.

Platythelys querceticola (Lindley) Garay

low ground orchid, jug orchid

Louisiana (east to) Florida; possibly extending to the Bahamas, West Indies, Mexico, Central America, and northern South America
Louisiana, Florida: rare and local in Florida and disjunct in central Louisiana; reported from Mississippi and Alabama
Plant: terrestrial, to 5–15 cm tall
Leaves: 3–8; cauline, ovate 1.5–4.0 cm wide × 5–8 cm long
Flowers: 8–25; arranged in a loose to dense terminal raceme; sepals ovate; petals rhomboidal, enclosed within the dorsal sepal forming a hood; lip oblong, constricted above the tip; the apex 3-lobed with the central lobe cordate and recurved; perianth whitish-green; individual flower size 3–4 mm with a saccate spur; capsule 5.5–6.0 mm, prominently ribbed
Habitat: swamps, river runs, and shaded floodplains
Flowering period: late July–September

Although many records have been documented for this species, especially in Florida, few extant sites appear to remain. Despite the fact that the delicate little white spikes of the low ground orchid may stand out dramatically in the dark, rich floodplains of the river bottoms, plants are often hidden under the fronds of near by ferns. In several of the known sites the plants of *Platythelys* are accompanied by the **fragrant ladies'-tresses,** *Spiranthes odorata,* **shadow-witch,** *Ponthieva racemosa,* and, in Florida, the **Florida adder's-mouth,** *Malaxis spicata.* Regular reports from Mississippi and Alabama indicate that this species is most likely present there as well and these records need to be documented. Plants may be overlooked and the habitat is not the most hospitable in early August when the species is in flower!

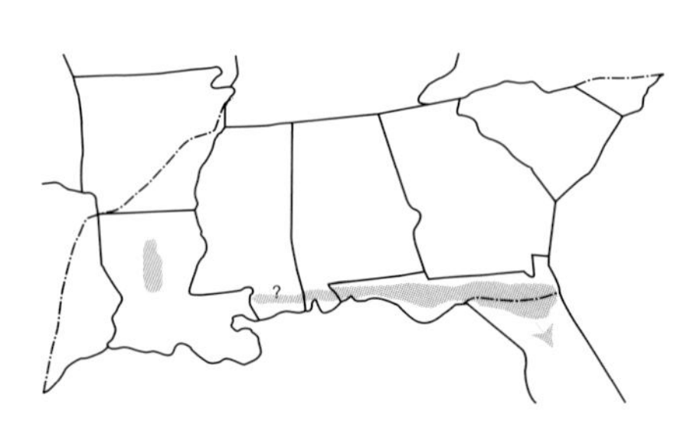

Pogonia

Pogonia is a small genus of only three species that is found in both Asia and North America. Formerly the genus included additional species that are now treated in *Triphora*, *Isotria*, and *Cleistes*, although some current authors are again including *Isotria* and *Cleistes* within *Pogonia*. We have only a single species in the United States and Canada, *P. ophioglossoides*, which has one of the broadest ranges of any North American orchid.

Pogonia ophioglossoides Ker-Gawler

rose pogonia; snakemouth orchid

Manitoba east to Newfoundland, south to Texas and Florida

forma *albiflora* Rand & Redfield—white-flowered form

Flora of Mount Desert Island 152. 1894, type: Maine

forma *brachypogon* (Fernald) P. M. Brown—short-bearded form

Rhodora 23: 245. 1922, as *Pogonia ophioglossoides* var. *brachypogon*, type: Nova Scotia

Texas, Louisiana, Arkansas, Alabama, Mississippi, Georgia, Florida, South Carolina, North Carolina: locally abundant in all states
Plant: terrestrial, 8–35 cm tall
Leaves: 1, rarely 2; cauline, ovate, placed midway on the stem, 2 cm wide × 6–8 cm long; a smaller leaflike bract subtends the flower
Flowers: 1–3(4); terminal; sepals and petals similar, lanceolate; the sepals wide spreading; lip spatulate with a deeply fringed margin and bright yellow beard or, in the forma *brachypogon*, the beard reduced to a few colorless knobs, to 2 cm; perianth from light to dark, rosy-pink or lavender or, in the forma *albiflora*, pure white; individual flower size ca. 4 cm
Habitat: moist meadows, open bogs and prairies, roadside ditches, and sphagnous seeps
Flowering period: March–May

From Newfoundland to Florida and westward to the Mississippi Valley, this little jewel adorns open bogs and meadows, roadside ditches, borrow pits, and sphagnous seeps. In Florida its pink flowers and broad, spatulate lip may remind one of roseate spoonbills. Although variable in color, form, and size, the rose pogonias are a true herald of a solid spring of wild orchids. They are often seen in the company of grass-pinks, *Calopogon*, ladies'-tresses, *Spiranthes*, and several of the carnivores such as pitcher-plants, sundews, and butterworts. Color and form vary greatly from colony to colony. It is not unusual to find plants with petals and sepals very narrow and, within the same colony, individuals with sepals and petals broad and rounded. Plants can have coloring from pale lilac to intense magenta and occasionally have white-flowered plants among them. Although most plants have solitary flowers, it can be interesting, especially in a large population, to search for those with multiple flowers. Stems with two flowers are not that unusual and, upon rare occasion, a three-flowered stem will be seen. Only once has the author seen a stem with four flowers! The unusual forma *brachypogon*, although described originally from Nova Scotia, has recently been seen, or more correctly detected, throughout the range of the species.

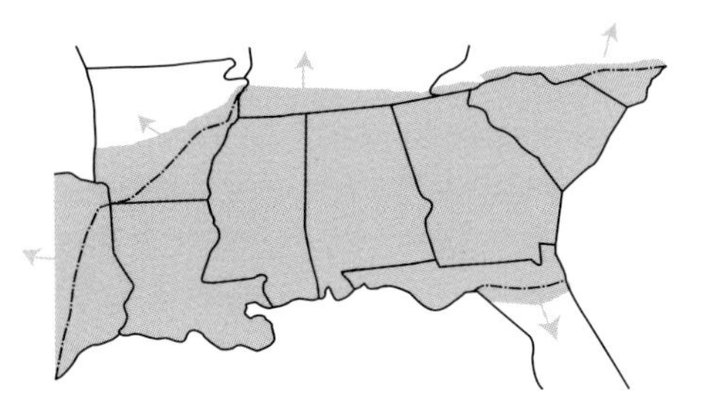

forma *albiflora*

Ponthieva

Ponthieva is a genus of about 50 neotropical species, which form herbaceous basal rosettes and a pubescent to puberulent raceme of flowers. The individual flowers are usually set at right angles to the axis. Two species are known from the United States and one, *P. racemosa*, from the Southeast.

Ponthieva racemosa (Walter) C. Mohr

shadow-witch

southeastern Virginia south to Florida, and west to eastern Texas; West Indies, Mexico, Central America, northern South America
Texas, Louisiana, Alabama, Mississippi, Georgia, Florida, South Carolina, North Carolina: local to frequent throughout its range, primarily within the Coastal Plain
Plant: terrestrial, 8–60 cm tall
Leaves: 3–8, in a basal rosette; dark green, elliptic, 1–5 cm wide × 3–15 cm long
Flowers: 8–30, non-resupinate; sepals light green, veined with darker green; petals white, veined with bright green; lip white with a green, concave center; individual flower size (6)8–(9)12 mm, the flowers tipped out from the rachis at about 60° and the ovary brown in color
Habitat: damp to wet shaded woodlands, swamps, and riverbanks
Flowering period: September–January

Ponthieva racemosa can often form large patches in wet woods along the seasonally flooded streambanks. Rarely is it found without other orchids and most often is accompanied by the **fragrant ladies'-tresses**, *Spiranthes odorata*, **Florida adder's-mouth**, *Malaxis spicata* (along the Atlantic Coastal Plain) and, in a few areas, the **low ground orchid**, *Platythelys querceticola*, and **southern oval ladies'-tresses**, *Spiranthes ovalis* var. *ovalis*.

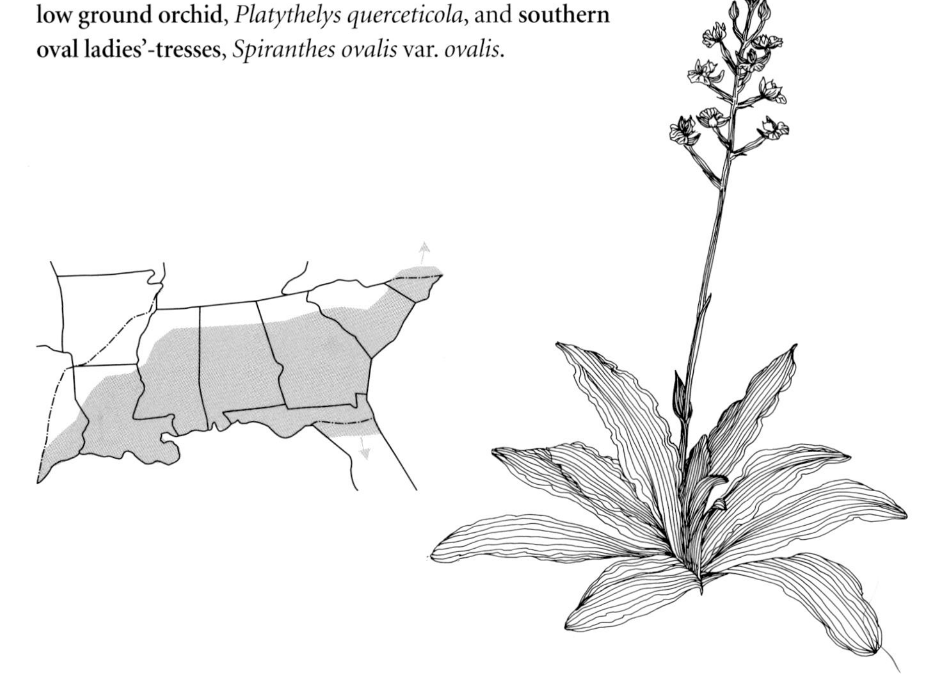

Pteroglossaspis

Another genus with strong African affinities, *Pteroglossaspis* consists of seven species, two that occur in the New World and one, the crestless plume orchid, *P. ecristata*, that can be found in the southeastern United States. A short spike of hooded flowers terminates the tall scapes and the leaves grow apart from the scape. The genus was formerly included within *Eulophia*.

Pteroglossaspis ecristata (Fernald) Rolfe

crestless plume orchid

North Carolina south to Florida and west to Louisiana; Cuba

forma *flava* P. M. Brown—yellow-flowered form

North American Native Orchid Journal 6(1): 64. 2000, type: Florida

forma *purpurea* P. M. Brown—dusky purple-flowered form

North American Native Orchid Journal 9: 34. 2003, type: South Carolina

Louisiana, Alabama, Mississippi, Georgia, Florida, South Carolina, North Carolina: very rare and local on the Coastal Plain; not appearing with regularity each year

Plant: terrestrial, 50–170 cm tall

Leaves: 3–4; lanceolate, yellow-green, up to 70 × 3.5 cm

Flowers: 10–30; the inflorescence axillary with the flowers in a terminal raceme atop an elongated scape; sepals and petals similar, lanceolate, lemon-yellow to grass-green, forming an elaborate hood; lip 3-lobed, ovate purple-brown to black with a green-yellow margin or, in the forma *flava*, the perianth entirely yellow or, in the forma *purpurea*, the perianth entirely shades of purple; individual flower size ca.1.5–2.0 cm

Habitat: old fields, orchards, pine flatwoods, prairies; usually in sandy soils

Flowering period: August–October

The crestless plume orchid is perhaps one of the least orchid-like species we have in the region, and was, at one time, a fairly frequent species to be found throughout much of the Gulf and Atlantic Coastal Plain. In recent times it has declined dramatically; coupled with the fact that the plants do not appear every year, it is now very difficult to find. The tall scapes produce a relatively few-flowered raceme of green and black orchids on a stick. The yellow-flowered form, forma *flava*, has been seen in Florida and Alabama. Throughout South Carolina and Louisiana the purple-flowered form, forma *purpurea,* appears to dominate. Colonies entirely of this color are not unusual.

forma *purpurea*

forma *lutea*

Sacoila

Sacoila is the showiest and largest-flowered genus among the many generic *Spiranthes* segregates. Although there are only ten species in the genus, it is exceedingly widespread throughout the New World tropics. A single species is found in the southeastern United States, and that in Florida. The genus *Sacoila* can be distinguished from the closely allied *Stenorrhynchos*, which does not occur in the Southeast, by the presence of a mentum, or saccate spur, at the base of the flower.

Sacoila lanceolata (Aublet) Garay var. *lanceolata*

leafless beaked orchid

Florida; West Indies, Mexico, Central America, South America
forma *albidaviridis* Catling & Sheviak—white and green-flowered form
Lindleyana 8(2): 77–81. 1993, type: Florida
forma *folsomii* P. M. Brown—golden bronze-flowered form
North American Native Orchid Journal 5(2): 169–73. 1999, type: Florida

Florida: a single record from Walton County in the Panhandle
Plant: terrestrial, 20–60 cm tall
Leaves: 4–6; oblanceolate, 2–8 cm wide × 5–35 cm long; absent at flowering time
Flowers: 10–40; in a terminal raceme atop a stout scape; sepals and petals similar, lanceolate, coral to brick-red; lip lanceolate, tapered at the apex, white to pale pink or, in the forma *albidaviridis*, the perianth white and green; or, in the forma *folsomii*, the perianth golden with a rose flush; individual flower size ca. 2 cm
Habitat: road shoulders and median strips along the highways, old fields, pine flatwoods
Flowering period: April–June

The leafless beaked orchid, or scarlet ladies'-tresses, is one of the most striking of all the southern orchids. The inclusion of *Sacoila lanceolata* in this work is based on a 1986 collection by Ruben Sauleda from Walton County, Florida. This location is a great distance north and/or west of the nearest known population. The two color forms are usually mixed in with the typical red, but given the small size of this disjunct site neither has been seen. The nomenclature of this species is perhaps the most convoluted of any orchid we have. Luer (1972) lists 34 synonyms in eight different genera! Brown (2000) gives the full history of this species. *Sacoila lanceolata* var. *paludicola* is restricted to southernmost Florida.

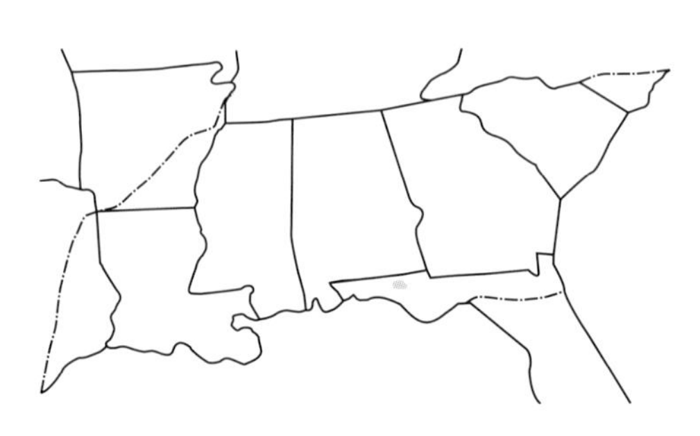

Spiranthes

Spiranthes is a cosmopolitan genus of about 50 species. Treated in the strictest sense, it is one of the most easily recognized genera but includes some of the more difficult plants to identify to species. The relatively slender, often twisted stems and spikes of small white or creamy-yellow (or pink in *S. sinensis*) flowers are universally recognizable. In the United States and Canada we have 24 species, including 17 that can be found in the Southeast. Plants of *Spiranthes torta* (Thunberg) Garay & Sweet, recorded in the literature for Louisiana, have proven to be *S. eatonii.*

Note: *Spiranthes cernua* is a compilospecies, with gene flow from several others species, depending on the plant's geographic location. Occasionally these plants prove problematic to key out. An unusual, nearly yellow-flowered cleistogamous/peloric race occurs in eastern Texas and western Louisiana.

1a leaves somewhat ascending the stem and relatively narrow and grass-like at flowering time . . . 2
1b leaves otherwise: basal, oval, orbicular, oblanceolate, or absent . . . 13

2a plants essentially spring flowering (March–May) . . . 3
2b plants essentially summer or autumn flowering (June–November) . . . 6

3a flowers white to ivory . . . 4
3b flowers green to creamy green . . . **woodland ladies'-tresses**, *S. sylvatica*

4a sepals appressed . . . **giant ladies'-tresses**, *S. praecox*
4b sepals spreading . . . 5

5a hairs pointed; late winter to spring flowering . . . **grass-leaved ladies'-tresses**, *S. vernalis*
5b hairs ball-tipped; late spring to summer flowering . . . **lace-lipped ladies'-tresses**, *S. laciniata*

6a plants essentially late spring to early summer flowering (June–July) . . . 7
6b plants essentially autumn flowering (September–November) . . . 12

7a lip creamy yellow; usually contrasting with petals and sepals; spring–summer flowering . . . 8
7b lip often creamy yellow or green; usually contrasting with petals and sepals; autumn flowering . . . 9

8a hairs long-pointed; plants of a variety of habitats, especially lawns and roadsides, rarely in standing water . . . **grass-leaved ladies'-tresses**, *S. vernalis*
8b hairs ball-tipped, plants primarily of wet areas, roadside ditches and streamsides, seasonally flooded savannas . . . **lace-lipped ladies'-tresses**, *S. laciniata*

9a plants stoloniferous, usually in damp to wet areas in both sun and shade; inflorescence a dense spike; lip often yellow or greenish . . . **fragrant ladies'-tresses,** S. *odorata*

9b plants otherwise . . . 10

10a plants of rich woodlands and second-growth forests; a cauline leaf prominent . . . 11

10b plants of various habitats, especially fields, meadows, and glades . . . 12

11a flowers fully sexual, rostellum present; all flowers opening fully and fruiting sequentially. . . **southern oval ladies'-tresses,** S. *ovalis* var. *ovalis*

11b flowers lack a rostellum, therefore self-pollinating; frequently not all flowers opening fully; flowers fruiting simultaneously . . . **northern oval ladies'-tresses,** S. *ovalis* var. *erostellata*

12a margins of the lateral sepals distinctly separated from the dorsal sepal, flowers creamy-white to nearly straw-colored, with the underside of the lip often a rich butterscotch color . . . **yellow ladies'-tresses,** S. *ochroleuca*

12b margins of the lateral sepals clearly touching or approximate to the dorsal sepal; flowers milky white to creamy-ivory, the center of the lip rarely pale ivory or greenish . . . **nodding ladies'-tresses,** S. *cernua*

13a plants spring flowering . . . 14

13b plants summer and/or autumn flowering. . .16

14a flowers white with a green throat . . . **Eaton's ladies'-tresses,** S. *eatonii*

14b flowers creamy yellow. . .15

15a inflorescence essentially glabrous . . . **Florida ladies'-tresses,** S. *floridana*

15b inflorescence densely pubescent . . . **short-lipped ladies'-tresses,** S. *brevilabris*

16a plants essentially summer flowering (June–August) . . . 17

16b plants essentially autumn flowering (September–November) . . . 18

17a flowers pure, pristine white . . . **little ladies'-tresses,** S. *tuberosa*

17b flowers with a green central portion on the lip . . . **southern slender ladies'-tresses,** S. *lacera* var. *gracilis*

18a flowers appearing only partially open; petals with a longitudinal green stripe; plants of East Texas . . . **Navasota ladies'-tresses,** S. *parksii*

18b flowers otherwise . . . 19

19a flowers entirely white or cream; sepals arching; plants of prairie habitats . . . **Great Plains ladies'-tresses,** S. *magnicamporum*

19b flowers white, lip contrasting creamy-yellow in color; sepals divergent; plants of pine flatwoods and damp roadsides . . . **long-lipped ladies'-tresses,** S. *longilabris*

Spiranthes brevilabris Lindley

short-lipped ladies'-tresses, Texas ladies'-tresses

Florida west to eastern Texas
Texas, Louisiana, Alabama, Mississippi, Georgia, Florida: very rare, with most historical populations extirpated
Globally Threatened
Plant: terrestrial, 20–40 cm tall; densely pubescent with capitate hairs
Leaves: 3–6; ovate, 1–2 cm wide × 2–6 cm long, yellow-green; withering at flowering time
Flowers: 10–35; in a single rank, spiraled or secund; sepals and petals similar, elliptic; perianth ivory to yellow with a dense pubescence; lip oblong, with the apex undulate-lacerate; individual flower size ca. 4–5 mm
Habitat: grassy roadsides, cemeteries
Flowering period: late February–April

The delicate, nearly ephemeral short-lipped ladies'-tresses is one of four species that produce winter rosettes. The leaves on *Spiranthes brevilabris* are more of a yellow-green than the other three species. When traveling by auto it is difficult to pick out the creamy-yellow, densely pubescent flowers on the roadsides, whereas the white-flowered species tend to stand out regardless of their size. After several years of intensive fieldwork only a single site has been found, and that in peninsular Florida. Elsewhere in the southeastern United States no other extant populations could be found.

Spiranthes cernua (Linnaeus) L. C. Richard

nodding ladies'-tresses

South Dakota east to Nova Scotia, south to Texas and Florida
Texas, Louisiana, Arkansas, Mississippi, Alabama, Florida, Georgia, South Carolina, North Carolina: widespread and often frequent in all but Florida and Louisiana
Plant: terrestrial, 10–50 cm tall
Leaves: 3–5; appearing basal or on the lower portion of the stem; linear-oblanceolate, up to 2 cm wide × 26 cm long; ascending to spreading; leaves are usually present at anthesis in most races, although in the Deep South and some prairie races they are absent
Flowers: 10–50; in a spike, tightly to loosely spiraled, with 5 or more flowers per cycle, nodding from the base of the perianth or rarely ascending; sepals and petals similar, lanceolate; perianth white, ivory, or in some races greenish; lip oblong, broad at the apex, the central portion of the lip, in some races, creamy-yellow or green; the sepals approximate and extending forward, sometimes arching above the flower; individual flower size 0.6–10.5 mm
Habitat: wet to dry open sites, lightly wooded areas, moist grassy roadsides, pine flatwoods, etc.
Flowering period: September–November (December)

Of all of our native orchids in North America, *Spiranthes cernua* is the most difficult to describe simply and concisely. Because it is a compilospecies—one that has gene flow from several different similar species—individuals in different geographic areas have strong resemblances to the basic diploid species contributing that area's unidirectional gene flow. In other words, those plants growing along the Atlantic Coastal Plain in close proximity to *S. odorata* have a greater resemblance to *S. odorata*, whereas those from the drier western portions of the range bear strong affinities to *S. magnicamporum*. Plants found in the northern states, primarily beyond the Southeast, would have gene flow from *S. ochroleuca*. In many areas, especially away from the Coastal Plain, this is the common autumn flowering *Spiranthes*. Those who are seriously interested in learning and determining the various races in their areas are urged to read Sheviak's entries for *S. cernua, S. odorata, S. magnicamporum*, and *S. ochroleuca* in volume 26 of *Flora of North America* (2002).

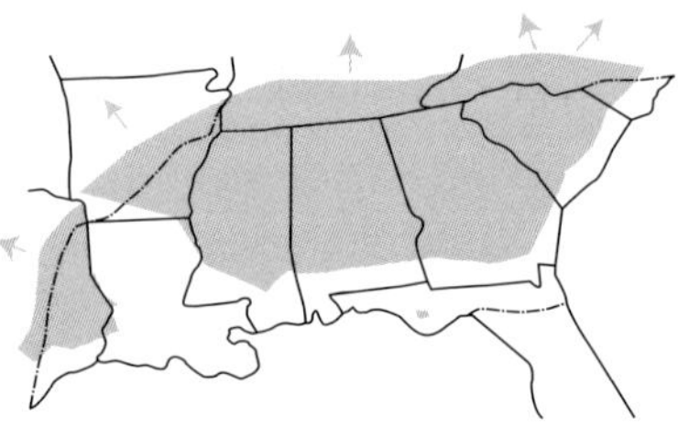

Perhaps the most interesting of the many regional races that occur in the Southeast is the recently found Deep South race, previously known from a few herbarium specimens collected along the Gulf coast and now from an extant site in the Apalachicola National Forest in the Panhandle of Florida. These plants flower very late, into December, are essentially leafless, and have very small, 5–7 mm flowers. In many ways they are unlike any other *Spiranthes* in our region. The late flowering time and small flower size may indicate gene flow from *Spiranthes ovalis*. Most plants of *S. cernua* seen throughout the Southeast are fairly similar to each other. Only in areas where the monoembryonic diploid species occur will there be plants that strongly emphasize the unidirectional gene flow.

cliestogamous race

Deep South Race

Spiranthes eatonii Ames *ex* P. M. Brown

Eaton's ladies'-tresses

eastern Texas east through all of Florida, north to southeastern Virginia; restricted to the Coastal Plain

Texas, Louisiana, Alabama, Mississippi, Georgia, Florida, South Carolina, North Carolina: apparently rare and local, although there are numerous herbarium vouchers; perhaps overlooked

Plant: terrestrial, 20–50 cm tall

Leaves: 3–6; oblanceolate-lanceolate, 0.75–1.0 cm wide × 5.5 cm long, withering quickly at flowering time

Flowers: 10–35; in a single rank, spiraled or secund; sepals spatulate, green at the base; petals lanceolate, green at the base; perianth white; lip oblong, centrally green with the apex undulate; individual flower size ca. 4–5 mm

Habitat: roadsides, cemeteries, drier pine flatwoods, sandy openings, damp saw palmetto scrub

Flowering period: late March–early May

In the examination of herbarium specimens this recently described species, *Spiranthes eatonii* (Brown, 1999), has been easily confused with *Spiranthes lacera* var. *lacera* and var. *gracilis*, *S. floridana*, *S. brevilabris*, *S. tuberosa*, and *S. torta.* In the field it is easily identified as it is the only spring-flowering, white-flowered, basal-leaved *Spiranthes* to be found within the Coastal Plain. The narrow, oblanceolate leaves are distinctive within this basal-leaved group. For many years plants of *S. eatonii* were identified as *S. lacera* and its southern variety *gracilis*, but neither name is a synonym. The exceedingly slender stems and tiny flowers make it easy to overlook when in bloom.

Spiranthes floridana (Wherry) Cory *emend.* P. M. Brown

Florida ladies'-tresses

eastern Texas east through all of Florida and north to North Carolina, primarily on the Coastal Plain

Texas, Louisiana, Alabama, Mississippi, Georgia, Florida, South Carolina, North Carolina: very rare and local, with many of the vouchered localities extirpated; recently confirmed populations are known only from Mississippi and Florida

Globally Threatened

Plant: terrestrial, 20–40 cm tall; glabrous to sparsely pubescent

Leaves: 3–5; ovate, 1–2 cm wide × 2–6 cm long, yellow-green, withering at flowering time

Flowers: 10–35; in a single rank, spiraled or secund; sepals and petals similar; perianth creamy-yellow; lip oblong, centrally yellow with the apex undulate; individual flower size ca. 4–5 mm

Habitat: roadsides, cemeteries, pine flatwoods

Flowering period: late March–early May

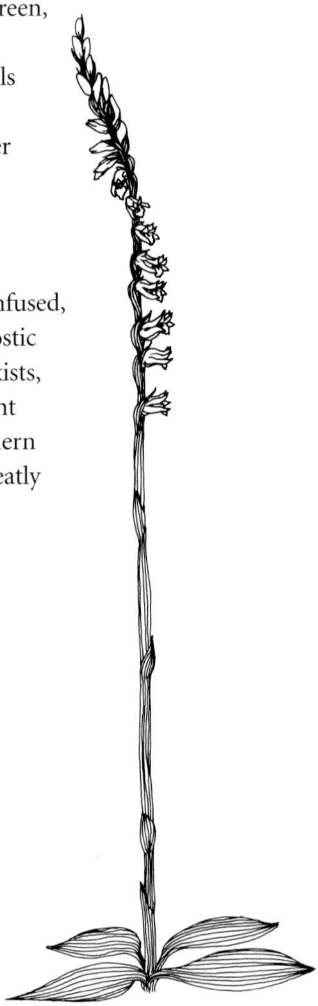

Spiranthes floridana and *S. brevilabris* are often easily confused, although the degree of pubescence is an excellent diagnostic tool in the field. While abundant apparent habitat still exists, *S. floridana* species has become very rare, with only recent populations seen in Mississippi (Sorrie, 1998) and northern Florida. Plants tend to be small and populations vary greatly from year to year.

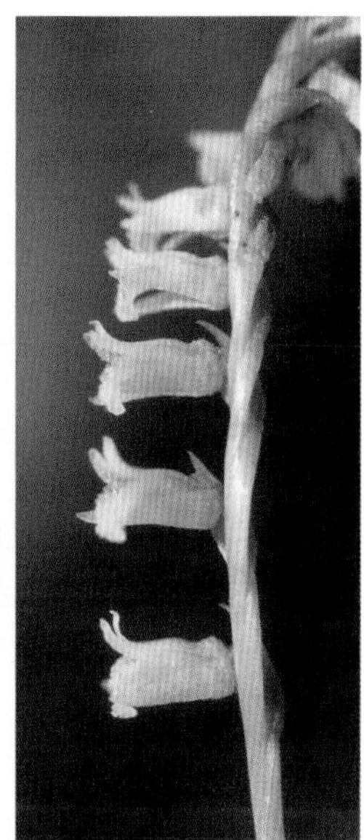

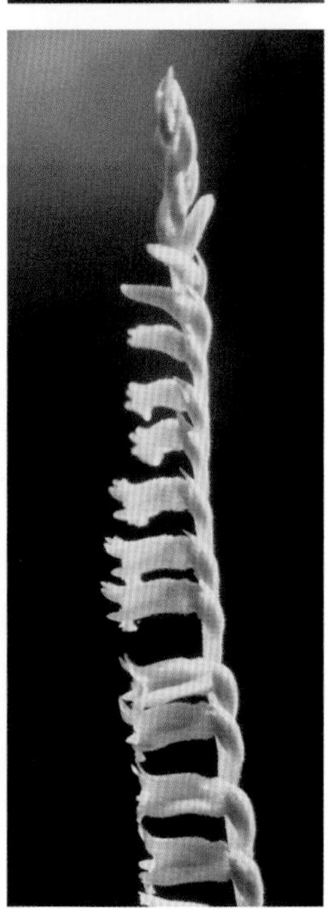

Spiranthes lacera (Rafinesque) Rafinesque var. *gracilis* (Bigelow) Luer

southern slender ladies'-tresses

Michigan east to Maine, south to Kansas, Texas, and Georgia
Texas, Louisiana, Arkansas, Alabama, Mississippi, Georgia, South Carolina, North Carolina: widespread and relatively common away from the Coastal Plain
Plant: terrestrial, 15–65 cm tall; glabrous to sparsely pubescent
Leaves: 2–4; ovate, dark green, 1–2 cm wide × 2–5 cm long, usually absent at flowering time
Flowers: 10–35; in a single rank, spiraled or secund; sepals and petals similar, elliptic; perianth white; lip oblong, with the apex rounded; central portion green with a clearly defined crisp apron; individual flower size 4.0–7.5 mm
Habitat: dry to moist meadows, grassy roadsides, cemeteries, open sandy areas in woodlands, lawns, old fields
Flowering period: July–September

Spiranthes lacera var. *gracilis* is the more southerly of the two varieties and found in nearly every state in the central and eastern United States. The small, white, green-throated flowers are very distinctive and the simple spiral of the inflorescence quite eye-catching. Plants are often encountered in lawns, and local cemeteries are usually a good place to search as well. The differences between this variety and the more northerly var. *lacera* are not great, but the more northern of the two has the lower flowers well spaced out on the inflorescence and they appear to be much smaller because of the position of the sepals.

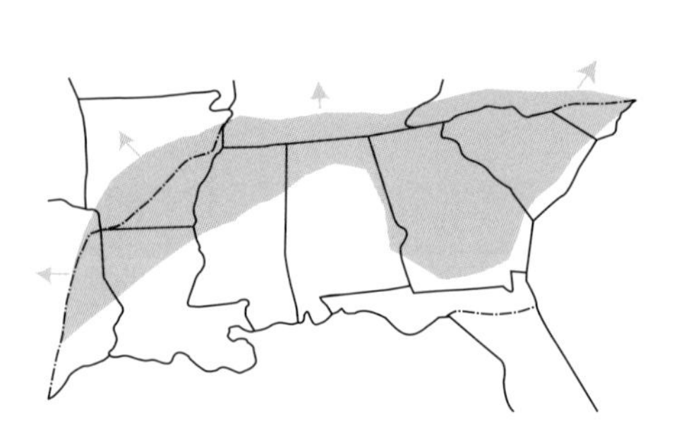

Spiranthes laciniata (Small) Ames

lace-lipped ladies'-tresses

eastern Texas east through all of Florida and north to southern New Jersey; primarily on the Coastal Plain

Texas, Louisiana, Alabama, Mississippi, Georgia, Florida, South Carolina, North Carolina: widely scattered and locally frequent, primarily along the Coastal Plain

Plant: terrestrial, 20–95 cm tall; densely pubescent with capitate hairs

Leaves: 3–5; lanceolate, 1.0–1.7 cm wide × 5–40 cm long

Flowers: 10–50; in a single rank, spiraled or secund; sepals and petals similar, elliptic; perianth white to ivory; lip oblong, with the apex undulate-lacerate, the central portion of the lip creamy yellow; individual flower size ca. 1 cm

Habitat: wet, grassy roadsides, ditches, swamps, and shallow open water

Flowering period: May–July

Spiranthes laciniata is easily distinguished from *S. vernalis*, which it superficially resembles, by its ball-tipped hairs, which differ from the pointed articulate hairs found on *S. vernalis*. It typically flowers later than *S. vernalis* where the two are sympatric. The tall, creamy white spikes may be a frequent sight along the wet roadside ditches and open savannas of the Gulf Coastal Plain.

Spiranthes longilabris Lindley

long-lipped ladies'-tresses

eastern Texas east through all of Florida, and north to southeastern Virginia
Texas, Louisiana, Alabama, Mississippi, Georgia, Florida, South Carolina, North Carolina: widely distributed but never frequent; most sites are local and with few plants
Plant: terrestrial, 20–50 cm tall; sparsely pubescent with clubbed hairs
Leaves: 3–5; linear-lanceolate, 0.5 cm wide × 8–15 cm long; often withered at flowering time
Flowers: 10–30; in a tight single rank, spiraled or secund; sepals and petals similar, lanceolate; perianth white to ivory; lip oblong, with the apex undulate-lacerate, the central portion of the lip yellow; the sepals wide-spreading; individual flower size 1.0–1.5 cm
Habitat: moist grassy roadsides, pine flatwoods
Flowering period: October–November

Spiranthes longilabris is perhaps the most handsome of all the ladies'-tresses. The close-ranked flowers spread their long sepals like wings and the narrow, creamy-yellow lip descends from the front of the large flowers. The long-lipped ladies'-tresses is an unmistakable plant and easily identified. It often grows with *S. odorata* and occasionally the hybrid, *S.* ×*folsomii* can also be found. Unfortunately, *S. longilabris* is becoming increasingly difficult to locate, especially in the southern portion of its range. Good, solid populations are still extant in the Green Swamp in southeastern North Carolina.

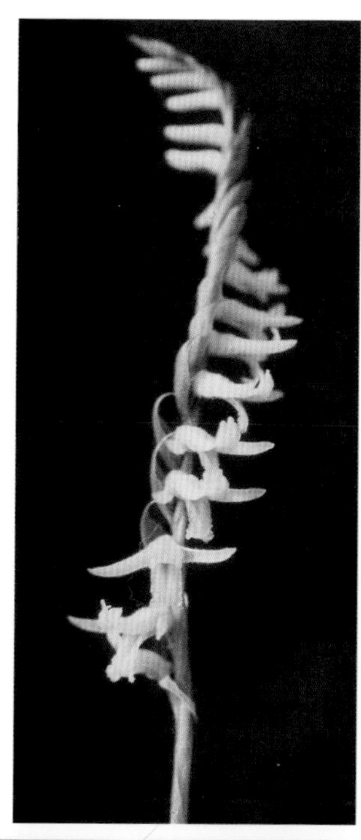

Spiranthes lucida (H. H. Eaton) Ames

shining ladies'-tresses

Wisconsin east to Nova Scotia, south to Kansas, Alabama, and West Virginia
Alabama: known only in the Southeast from a single site in the north-central part of the state
Plant: terrestrial, 10–38 cm tall
Leaves: 3–4; basal, elliptic-lanceolate, 0.5–1.5 cm wide × 3–12 cm long; the surface shiny; present at flowering time and long beyond
Flowers: 10–20; in an open spiral; nearly horizontal to nodding; sepals and petals similar, white, lanceolate; lateral sepals appressed to the petals and lip forming a tube; the lip oblong, bright yellow with the central portion yellow to orange or green, the margin crenulate; individual flower size 0.6–0.9 cm
Habitat: rocky riverbanks, seeps, fens; usually calcareous
Flowering period: May–June

The shining ladies'-tresses is unlike any other in the genus in North America. The broad, shining leaves and tubular yellow-lipped flowers set it apart. In the far northern portion of its range in North America the shining ladies'-tresses is often accompanied by plants of *Spiranthes romanzoffiana*, the **hooded ladies'-tresses**, a species that flowers in the summer or autumn. The plants in Alabama were recently discovered by Jim Allison (2001) and represent both the southern limit of the range and a disjunct population.

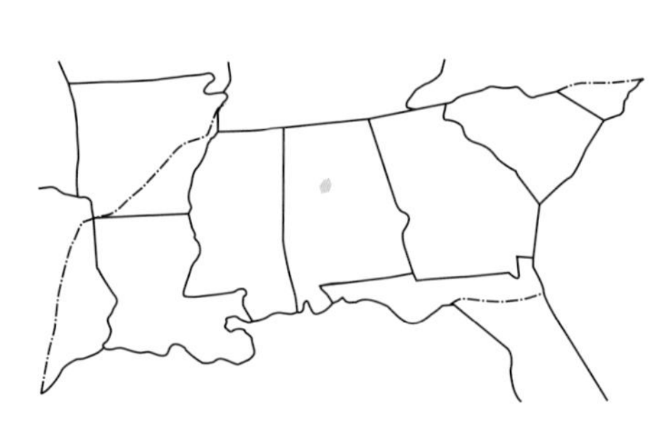

Spiranthes magnicamporum Sheviak

Great Plains ladies'-tresses

Manitoba east to southern Ontario, south to New Mexico, Texas, Pennsylvania, and Georgia

Texas, Louisiana, Arkansas, Alabama, Mississippi, Georgia: scattered disjunct locales in northern and central counties for this typical prairie species

Plant: terrestrial, 10–63 cm tall

Leaves: 2–4; appearing basal or on the lower portion of the stem; linear-oblanceolate, up to 1.5 cm wide × 16 cm long; ascending to spreading; leaves are usually absent or withering at anthesis

Flowers: 12–54; in a spike, tightly to loosely spiraled with 3–4 flowers per cycle; abruptly nodding from the base; lateral sepals wide spreading and usually arched above the flower, petals linear-lanceolate; perianth white, ivory, or cream; lip ovate to oblong, the apex crenulate, the central portion of the lip usually yellow; individual flower size 0.4–1.2 cm

Habitat: wet to dry prairies and fens

Flowering period: September–November

Spiranthes magnicamporum is a typical prairie species reaching the southeastern limit of its range in several widely scattered and nearly disjunct sites in the Southeast. In central Texas plants of *Spiranthes magnicamporum* are frequent, but do not reach the limits of the Southeast. This is one of the several species of *Spiranthes* that contributes gene flow to *S. cernua*. Several of these southeastern sites contain mixed populations of both *S. magnicamporum* and *S. cernua*. Careful examination of the plants and, if absolutely necessary, the seeds may be required to determine the correct species. Seeds from the basic diploid species, *S. magnicamporum*, *S. ochroleuca*, *S. odorata*, and *S. ovalis*, are monoembryonic whereas those of *S. cernua*, usually a polyploid species, are polyembryonic. This type of examination goes beyond the scope of this field guide and requires more sophisticated techniques. A compound microscope and the skills to use it are required.

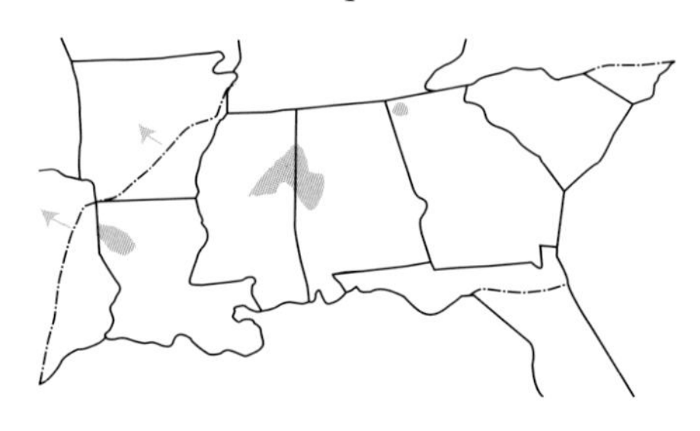

Spiranthes ochroleuca (Rydberg) Rydberg

yellow ladies'-tresses

Michigan east to Nova Scotia, south to Kentucky and South Carolina
South Carolina: known from only a single population north of Greenville
Plant: terrestrial, 10–55 cm tall
Leaves: 3–5; appearing basal or on the lower portion of the stem; linear-lanceolate, up to 2 cm wide × 21 cm long; ascending to spreading; leave are present at anthesis
Flowers: 10–50; in a spike, tightly to loosely spiraled with 3–4(5) flowers per cycle, ascending; sepals and petals similar, lanceolate; lateral sepals appressed to petals and lip, straight; perianth white to cream-colored; lip oblong to ovate, the central portion of the lip a deeper yellow, individual flower size 0.7–1.2 cm
Habitat: dry to somewhat moist open sites, ledges, barrens, slightly wooded areas, grassy roadsides
Flowering period: September–November

The inclusion of *Spiranthes ochroleuca* in the orchid flora of the Southeast is based on the identification of a specimen from Greenville County, South Carolina. The site still has extant plants, although they are located in an exceedingly difficult place to reach. Typically, *S. ochroleuca* has a distinct butterscotch-colored trough in the center of the lip that is very visible on the bottom side. The South Carolina plants superficially appear more like *S. cernua*, but when closely examined it is evident they are *S. ochroleuca*. The species is also found not far away in southwestern North Carolina.

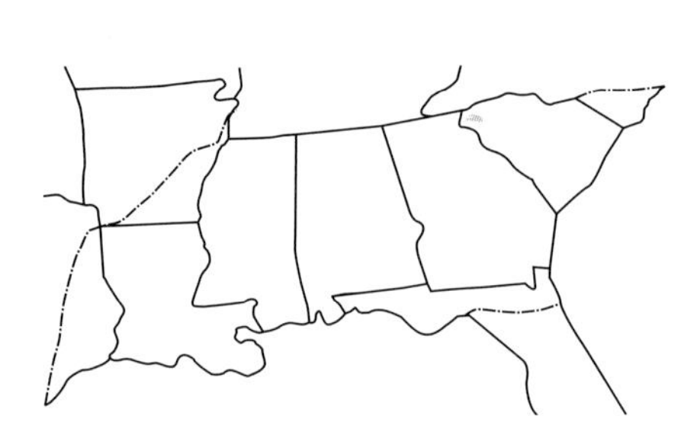

Spiranthes odorata (Nuttall) Lindley

fragrant ladies'-tresses

eastern Texas north to Oklahoma and Arkansas, east to Florida, and north to (?)Delaware

Texas, Louisiana, Arkansas, Alabama, Mississippi, Georgia, Florida, South Carolina, North Carolina: widespread and locally common through much of our region, especially on the Coastal Plain; disjunct sites occur in the central Appalachians and Kentucky

Plant: terrestrial or semiaquatic, 20–110 cm tall, pubescent, stoloniferous

Leaves: 3–5; linear-oblanceolate, up to 4 cm wide × 52 cm long; rigidly ascending or spreading

Flowers: 10–30; in several tight ranks; sepals and petals similar, lanceolate; perianth white to ivory; lip oblong, tapering to the apex, the central portion of lip creamy-yellow or green; the sepals extending forward; individual flower size 1.0–1.8 cm

Habitat: moist grassy roadsides, pine flatwoods, cypress swamps, wooded river floodplains

Flowering period: October–January

Spiranthes odorata can be by far the largest of our native ladies'-tresses. Plants in wooded swamplands can reach a full meter in height. The fragrant ladies'-tresses typically occurs in seasonally inundated sites and may bloom while emerging from shallow water. The rather thick, broad leaves give the plant a distinctive vegetative habit. The very long, wide-spreading roots produce vegetative offshoots often 30 cm from the parent shoot, giving rise to extensive clonal colonies. Despite the typical size of *Spiranthes odorata*, there is a very definable ecotype that occupies mowed road shoulders and can often be no more than 15 cm tall.

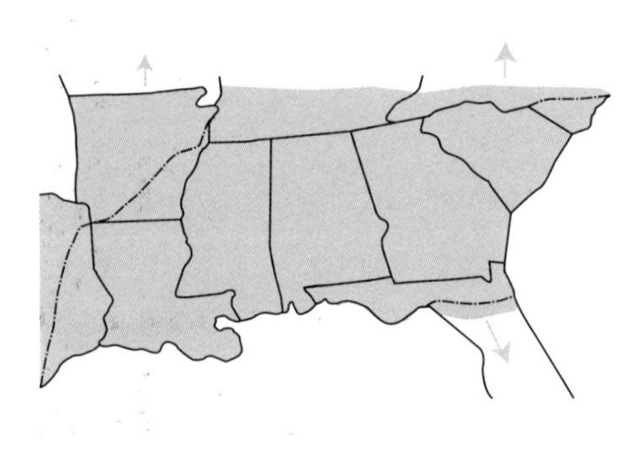

Dwarf roadside form.

Spiranthes ovalis Lindley var. *ovalis*

southern oval ladies'-tresses

Arkansas south to eastern Texas, and east to Florida
Texas, Louisiana, Arkansas, Alabama, Mississippi, Georgia, Florida: very rare and local with only a few populations in each state
Globally Threatened
Plant: terrestrial, 20–40 cm tall; pubescent
Leaves: 2–4; basal and on the lower half of the stem, oblanceolate, 0.5–1.5 cm wide × 3–15 cm long; present at flowering time
Flowers: 10–50; in 3 tight ranks; sepals and petals similar, lanceolate; perianth white; lip oblong, tapering to the apex with a delicate undulate margin, the sepals extending forward; individual flower size 5.5–7.0 mm; rostellum and viscidium present, therefore the plants sexual
Habitat: rich, damp woodlands and floodplains
Flowering period: October–November

One of the most charming of all the ladies'-tresses, *Spiranthes ovalis* is the only species with an exclusive woodland habitat. The pristine, small white flowers are usually carried in three distinctive, tight, vertical ranks. In the variety *ovalis* the flowers are sexually complete and therefore fertilization is effected by a pollinator. The flowers in var. *ovalis* are always fully expanded. This nominate variety is relatively rare and found only in a few states.

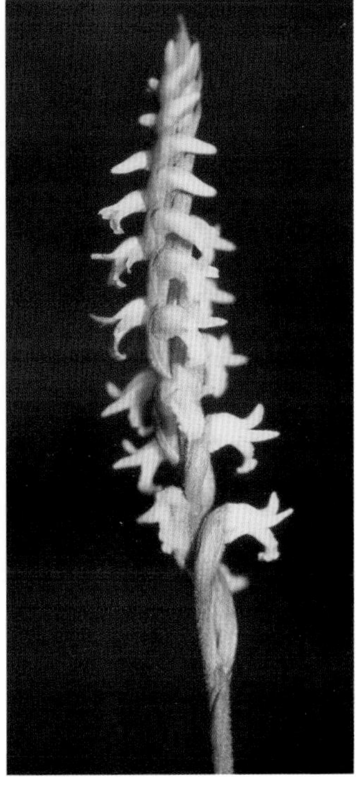

Spiranthes ovalis Lindley var. *erostellata* Catling

northern oval ladies'-tresses

Ontario south to Illinois and Arkansas, east to Ohio and western Pennsylvania, and south to northern Florida

Texas, Louisiana, Arkansas, Alabama, Mississippi, Georgia, Florida, South Carolina, North Carolina: wide-ranging and locally common especially at the northern limit of the Southeast; apparently becoming more common, especially in disturbed woodland sites

Plant: terrestrial, 20–40 cm tall, pubescent

Leaves: 2–4; basal and on the lower half of the stem, oblanceolate, 0.5–1.5 cm wide × 3–15 cm long, present at flowering time

Flowers: 10–50; in 3 tight ranks; sepals and petals similar, lanceolate; perianth white; lip oblong, tapering to the apex with a delicate undulate margin; the sepals extending forward; individual flower size 3.5–5.0 mm; rostellum lacking, therefore the plants are self-pollinating; flowers not always fully open

Habitat: rich, damp woodlands and floodplains

Flowering period: October–November

In most plants of *Spiranthes ovalis* var. *erostellata*, the flowers are never quite fully open, or in many individuals they are tiny cleistogamous flowers, and the ovaries are simultaneously swollen on each flower within the inflorescence. Those of *S. ovalis* var. *ovalis* have fully open flowers and the ovaries swell progressively. In a few sites in northern Florida both varieties grow together as well as *S. odorata*. The hybrid between them, *S.* ×*itchetuckneensis*, is often also present. For more information on hybrids within the genus *Spiranthes* see page 228.

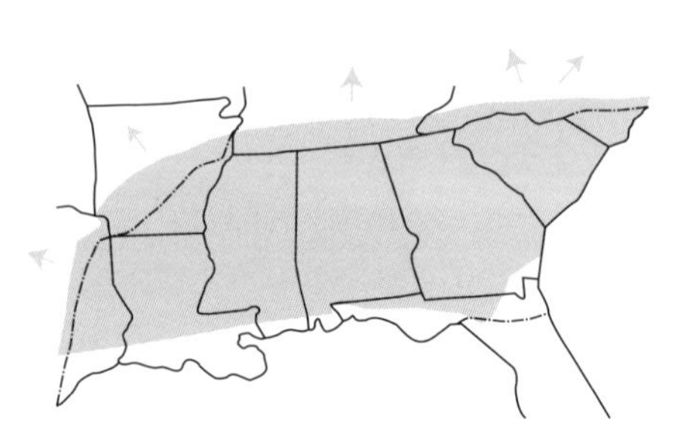

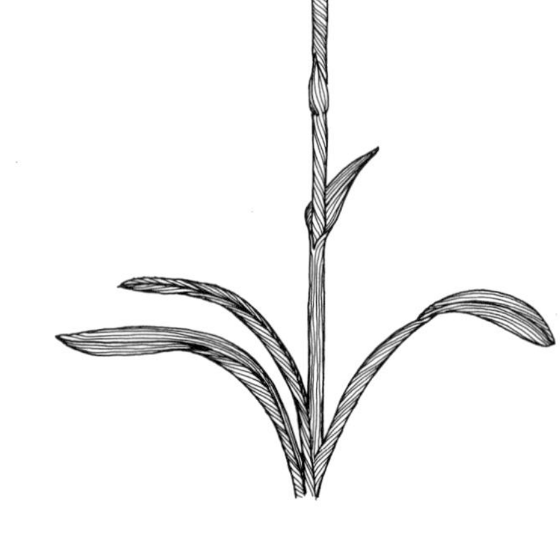

Spiranthes parksii Correll

Navasota ladies'-tresses

east-central Texas
FEDERALLY LISTED AS ENDANGERED
Texas: restricted to two areas within the Piney Woods and Post Oak regions
Plant: terrestrial, 15–33 cm tall, moderately to densely pubescent
Leaves: 2–3; appearing basal or on the lower portion of the stem; linear-lanceolate, up to 2 cm wide × 22 cm long; ascending to spreading; the leaves are usually absent or withering at anthesis
Flowers: 8–15; in a spike, loosely spiraled with 5 flowers per cycle; ascending; sepals lanceolate, white to pale green, directed forward; petals obovate, whitish to yellow-green with a prominent central green stripe, usually erose on the margin; lip oblong to obovate centrally yellowish or green, the margins erose at the apex; individual flower size 0.5–0.6 cm
Habitat: thin soil, lightly shaded woodlands usually with post oaks
Flowering period: October–November

Spiranthes parksii is one of the rarest ladies'-tresses in North America. It is confined to two areas in Texas, one of which falls within our range of the Southeast. Plants of this species were first described in 1947 and then rarely seen until Paul Catling rediscovered them in 1986. *Spiranthes parksii* was one of the very few species that Luer, in the early 1970s, was unable to locate to photograph for his book. Subsequently, in 1986, Bridges and Orzell found a small population in Angelina National Forest in eastern Texas. The tiny flowers never appear to completely open, or at least are not wide-spreading like most *Spiranthes*, and appear to be apomictic. The plants are not unlike some cleistogamous and peloric forms of *S. cernua* found in the area, but the small flower size and shape of the lateral sepals are distinctive. Like *S. cernua* the seeds appear to be polyembryonic.

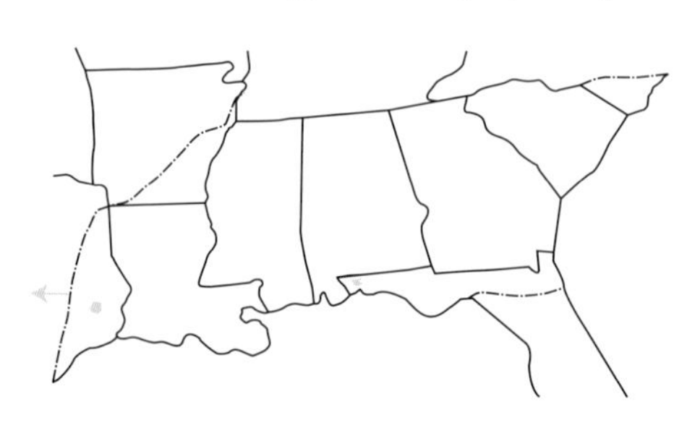

Spiranthes praecox (Walter) S. Watson

giant ladies'-tresses

Arkansas south to Texas, east to Florida, and north to New Jersey; primarily on the Coastal Plain

forma *albolabia* P. M. Brown & C. McCartney, white-lipped form
North American Native Orchid Journal 1(1): 13. 1995, type: Florida

Texas, Louisiana, (Arkansas), Alabama, Mississippi, Georgia, Florida, South Carolina, North Carolina: widespread and locally common, primarily along the Coastal Plain
Plant: terrestrial, 20–75 cm tall, sparsely pubescent
Leaves: 2–7; basal and on the lower half of the stem, linear-lanceolate, 1–7 cm wide × 10–25 cm long, present at flowering time
Flowers: 10–40; in either single or multiple ranks; sepals and petals similar, lanceolate; perianth typically white, but in some plants green; lip ovate-oblong, rounded to the apex with a delicate undulate margin with distinctive green veining or, in the forma *albolabia,* the lip appearing pure white and the veins actually a pale lemon yellow, the sepals appressed and extending forward to create a tubular flower; individual flower size 6–12 mm
Habitat: roadsides, meadows, prairies, open woodlands—just about anywhere that is not too shady
Flowering period: March–June

Spiranthes praecox is one of the most frequently encountered orchids in much of the Coastal Plain. It starts flowering shortly after *S. vernalis* and continues throughout the spring. The tubular flowers with their appressed sepals are distinctive and help to quickly separate it from *S. vernalis*. In rare situations the two hybridize, producing *S.* ×*meridionalis*. The small, slender, pure white flowers help in separating this species from the somewhat similar woodland ladies'-tress, *S. sylvatica*, which also has green veining in the lip.

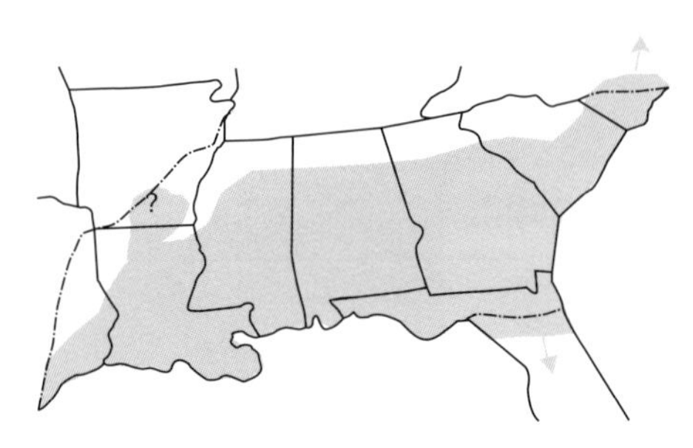

forma *albolabia*

Spiranthes sylvatica P. M. Brown

woodland ladies'-tresses

eastern Texas to Florida, north to Arkansas, and along the Coastal Plain to southeastern North Carolina

Texas, Louisiana, Arkansas, Alabama, Mississippi, Georgia, Florida, South Carolina, North Carolina: locally common in varied woodland habitats throughout the Coastal Plain region; recent reports (Slaughter, pers. comm.) indicate that all of the plants previously thought to be *Spiranthes praecox* in Arkansas are in actuality *S. sylvatica*.

Plant: terrestrial, 25–75 cm tall, sparsely pubescent with capitate hairs

Leaves: 3–7; basal and on the lower third of the stem, linear-lanceolate, 0.8–1.5 cm wide × 10–35 cm long, present at flowering time

Flowers: 10–30; in a dense spike usually appearing as multiple ranks; sepals and petals similar, lanceolate; perianth pale green to creamy-green; lip ovate-oblong, rounded and broadened to the apex with a delicate undulate or ruffled margin and with distinctive darker green veining; the sepals slightly spreading; individual flower size 1.0–1.7(2.2) cm

Habitat: shaded roadsides, open woodlands, and dry live oak hammocks; rarely in bordering wetlands

Flowering period: late March–early May

Spiranthes sylvatica is the most recent *Spiranthes* species to be described from North America (Brown, 2001). Although the plants have been known for some time, sufficient evidence has been available only recently to satisfactorily separate this species from *S. praecox*. Although both typically have green veined lips, all similarity ceases at that point. The woodland ladies'-tresses has been passed over for many years as a disappointing example of *S. praecox*. *Spiranthes sylvatica* is usually a plant of shaded and woodland habitats and the very distinctive large, creamy green flowers are unlike any other *Spiranthes*. It is most frequently seen along roadside hedgerows bordering woodlands, where the plants are tucked up under the shrubs into the border. Also, there are many other distinctive differences between the two species but flower size, shape, and color are the most noticeable. Recent fieldwork has confirmed extant sites for Texas, Arkansas, Georgia, Florida, South Carolina, and North Carolina.

Spiranthes tuberosa Rafinesque

little ladies'-tresses

Arkansas east to southern Michigan and Massachusetts, south to Florida, and west to Texas
Texas, Louisiana, Arkansas, Alabama, Mississippi, Georgia, Florida, South Carolina, North Carolina: a common summer-flowering species; primarily found along the Coastal Plain
Plant: terrestrial, 10–30 cm tall, glabrous
Leaves: 2–4; ovate, dark green, 1–2 cm wide × 2–5 cm long, absent at flowering time
Flowers: 10–35; in a single rank, spiraled or secund; sepals and petals similar, elliptic; perianth crystalline white; lip oblong, with the apex undulate-lacerate, exceeding the sepals; individual flower size ca. 3–4 mm
Habitat: grassy roadsides, cemeteries, open sandy areas in woodlands
Flowering period: June–July

Spiranthes tuberosa is the only pure white *Spiranthes* to flower in midsummer. One of its favorite habitats is old, dry cemeteries. The nomenclatural history of this plant is rather complex, and among the names applied to it are *S. beckii* and *S. grayi*. See Correll (1950) for a discussion. *S. tuberosa* can be easily recognized by its pure white flowers, broad, crisped lip, and the absence of leaves at flowering time. The dark green rosettes of leaves are prominent throughout the winter.

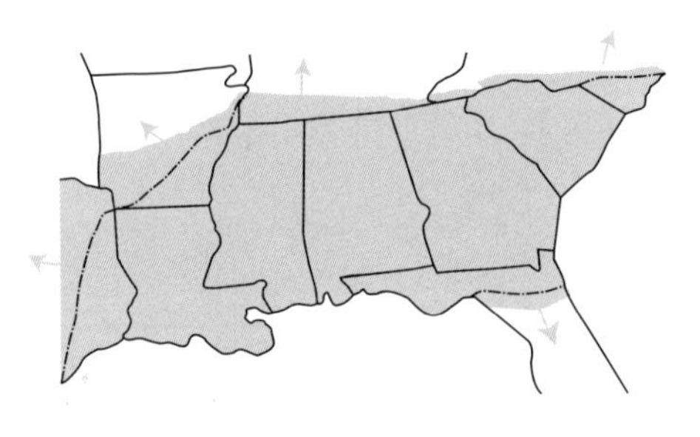

Spiranthes vernalis Englemann & Gray

grass-leaved ladies'-tresses

Nebraska south to Texas, east to Florida, and north to southern New Hampshire
Texas, Louisiana, Arkansas, Alabama, Mississippi, Georgia, Florida, South Carolina, North Carolina: the most common *Spiranthes* to be found in the southeastern United States; primarily along the Coastal Plain, but not unusual further inland
Plant: terrestrial, 10–85 cm tall, pubescent with sharp-pointed hairs
Leaves: 2–7; basal and on the lower third of the stem, linear-lanceolate, 1–2 cm wide × 5–25 cm long, present at flowering time
Flowers: 10–50; in either a single or multiple rank; sepals and petals similar, lanceolate; perianth typically creamy-white; lip ovate-oblong, usually a deeper creamy-yellow, rounded to the apex with a delicate, undulate margin; the sepals wide spreading; individual flower size 6–9 mm
Habitat: roadsides, meadows, prairies, untended lawns—just about anywhere that is sunny
Flowering period: March–May

Spiranthes vernalis is perhaps our most variable *Spiranthes* in habit, although the flowers remain surprisingly consistent. Plants may vary greatly in size and vigor as well as degree of spiraling, which results in plants essentially secund to those that appear to be multiple ranked. Plants are not consistent in habit from year to year. Color is somewhat variable, from nearly pure white to cream with a contrasting, more yellowy lip, and some individuals have two brown or orange spots on the lip. The most consistent diagnostic character is the presence in the inflorescence of copious articulate, pointed hairs that readily distinguish *S. vernalis* from other species of *Spiranthes*. It would be interesting to follow *S. vernalis* north with the flowering season, as it starts in late January and early February in south Florida and continues, month by month, until it reaches southern New England in late August and New Hampshire in early September—nearly a full year of one species of *Spiranthes*!

Hybrids:

Spiranthes ×*folsomii* P. M. Brown
(*S. longilabris* × *S. odorata*)
Folsom's hybrid ladies'-tresses
North American Native Orchid Journal. 6(1): 16. 2000, type: Florida

Plants tend to be intermediate between the parents and have shorter leaves than *Spiranthes odorata* and with more slender, wide-spreading sepals.

Spiranthes ×*intermedia* Ames
intermediate hybrid ladies'-tresses
(*S. lacera* var. *gracilis* × *S. vernalis*)
Rhodora 5: 262. 1903, type: Massachusetts

One of the first *Spiranthes* hybrids to be described, it only occurs in a few places, primarily north of the Southeast, where both parents flower at the same time.

Spiranthes ×*itchetuckneensis* P. M. Brown
(*S. odorata* × *S. ovalis* var. *ovalis*)
Ichetucknee hybrid ladies'-tresses

North American Native Orchid Journal 5(4): 362–66. 1999, type: Florida

This is a frequent hybrid where both parents are sympatric. They can be difficult to identify and look like either large *Spiranthes ovalis* or small *S. odorata*.

Spiranthes ×*meridionalis* P. M. Brown
(*S. praecox* × *S. vernalis*)
southern hybrid ladies'-tresses

North American Native Orchid Journal 6(2): 139. 2000, type: Florida

Despite the relative abundance of both parents and the fact that both often occupy the same habitat, relatively few specimens have been identified as this hybrid. Look for *Spiranthes vernalis* with appressed sepals and some green in the lip or *S. praecox* with divergent sepals.

Tipularia

Tipularia is a small genus of only two species known from the Himalayas and the eastern United States. The species are very similar to each other and differ in the shape of the lip. The genus is characterized by having a series of tubers from which arise a single, annual leaf that remains green throughout the winter, withering before the flower spike emerges in the summer. The leafless spike of flowers has the sepals and petals all drawn to one side. The single species in the Southeast is the crane-fly orchis, *T. discolor*.

Tipularia discolor (Pursh) Nuttall

crane-fly orchis

eastern Texas northeast to southern Michigan, and east to southeastern Massachusetts, and south to Florida

forma *viridifolia* P. M. Brown—green-leaved form

North American Native Orchid Journal (4): 336–37. 2000, type: Florida

Texas, Louisiana, Arkansas, Alabama, Mississippi, Georgia, Florida, South Carolina, North Carolina: common throughout most of the Southeast
Plant: terrestrial, 25–60 cm tall; the flowering stem brown
Leaves: 1; basal, ovate, dark green above with raised purple spots and dark purple beneath or, in the forma *viridifolia*, green on both sides; 6–7 cm wide × 8–10 cm, the long petiole ca. 5 cm; leaf withering in the spring and absent at flowering time
Flowers: 20–40; in a loose raceme; sepals and petals similar, oblanceolate; perianth greenish-yellow, tinged and mottled with pale purple; lip 3-lobed, the central lobe slender, blunt, with a few shallow teeth; the sepals and petals asymmetrical and all drawn to one side; individual flower size 2–3 × 3.0–3.5 cm not including the 1.5–2.2 cm spur
Habitat: deciduous and mixed woodlands
Flowering period: late June–mid-August

Plants of the very distinctive crane-fly orchis are most easily found during the winter months when not in flower, when the single, long-petioled leaf is most apparent. The dark green leaf has a seersucker look with raised purple spots on the upper surface. If one turns over the leaf the satiny purple underside can be seen—hence the name *discolor* or two colors. Plants with no purple present, forma *viridifolia*, have been seen in scattered locations. The flower spike appears in midsummer and the brown coloring of the stem and flowers makes it most difficult to see in the woodlands, although the spike and flowers are not at all small. Spikes usually grow to 45–50+ cm tall.

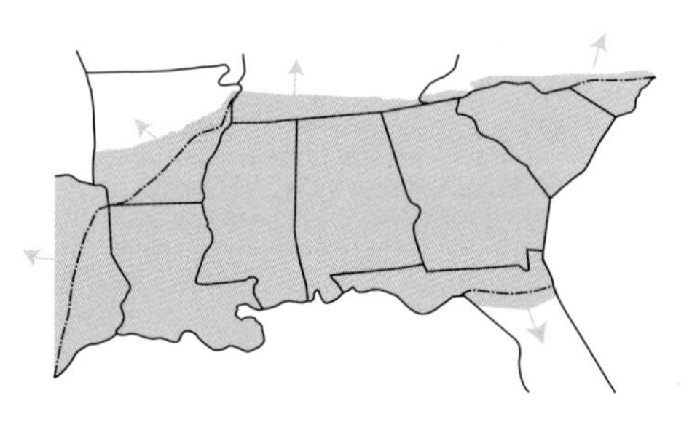

Triphora

The genus *Triphora* consists of about 20 species in North America, the West Indies, Mexico, and Central America. It is primarily a genus of small, delicate herbs, many of which may be largely mycotrophic. They all arise from swollen tuberoids and produce, in some, very colorful, although small, flowers. Several species have flowers that do not fully open. In the Southeast we have two species.

1a flowers nodding, white to pink to lavender, fully open; lip lowermost . . . **three birds orchid**, *T. trianthophora*

1b flowers upright, yellow, not fully open; lip uppermost . . . **Rickett's noddingcaps**, *T. rickettii*

Triphora rickettii Luer

Rickett's noddingcaps

Florida
Florida: restricted to a few sites in central Florida
Globally Threatened
Plant: terrestrial, 8–20 cm tall
Leaves: 5–10; broadly ovate-cordate, with undulate margins, dark green; 1–2 cm × 1–2 cm
Flowers: 1–8, erect; from the axils of the leaves; sepals and petals similar, oblanceolate; perianth yellow, flowers not opening fully; lip 3-lobed, the central lobe with the margin undulate and 3 parallel crests; individual flower size ca. 1 cm
Habitat: deciduous and mixed damp woodlands
Flowering period: late July–early October

The inclusion of Rickett's noddingcaps in the orchid flora of the Southeast is based on Luer's eyewitness report of plants in Columbia County, Florida (Luer, 1975). Although no specimen exists, his report is considered valid. *Triphora rickettii* was known for many years from central Florida before it was described by Luer in 1966. The tiny yellow flowers, which never fully open, are similar in appearance to *Triphora gentianoides*, but the leaves are totally different. They resemble the leaves of *T. craigheadii* in shape but are green on both sides. Some authors have reduced *T. rickettii* to synonymy with *T. yucatenensis*, but there are just too many differences, including color and position of the flowers, for that opinion to be acceptable.

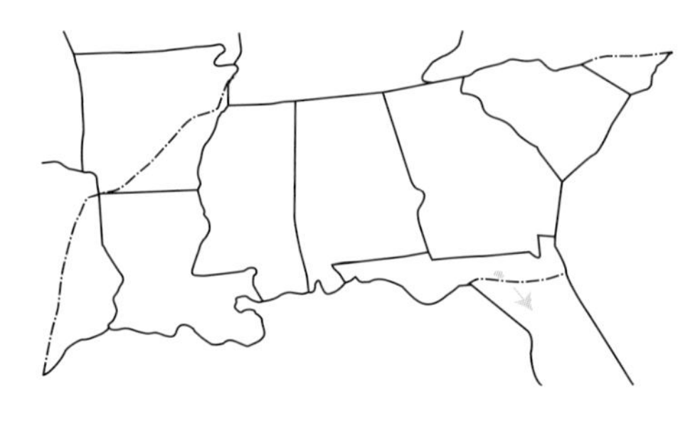

Triphora trianthophora (Swartz) Rydberg subsp. *trianthophora*

three birds orchid

Texas north to Minnesota, east to Maine, and south to Florida

forma *albidoflava* Keenan—white-flowered form

Rhodora 94: 38–39. 1992, type: New Hampshire

forma *caerulea* P. M. Brown—blue-flowered form

North American Native Orchid Journal 7(1): 94–95. 2001, type: Florida

forma *rossii* P. M. Brown—multi-colored form

North American Native Orchid Journal 5(1): 5. 1999, type: Florida

Texas, Louisiana, Arkansas, Alabama, Mississippi, Georgia, Florida, South Carolina, North Carolina: rare and local throughout

Plant: terrestrial, 8–25 cm tall

Leaves: 2–8; broadly ovate-cordate, with smooth margins, dark green often with a purple cast or, in the forma *rossii*, the stem and leaves white, pink, and yellow; 10–15 × 5–15 mm

Flowers: 1–8(12), nodding; from the axils of the upper leaves; sepals and petals similar, oblanceolate; perianth white to pink; lip 3-lobed, the central lobe with the margin sinuate and 3 parallel green crests or, in the forma *albidoflava*, the perianth pure white and the crests yellow or, in the forma *caerulea*, lilac-blue; individual flower size ca. 1–2 cm

Habitat: deciduous and mixed woodlands, often with partridgeberry

Flowering period: late July–mid-September

Triphora trianthophora is the largest-flowered and showiest member of the genus *Triphora* in the United States. The plants are quite elusive and appear for only a few days most years. The stunning little flowers open in midmorning and usually close by midafternoon, leaving only a few hours for the eager eye to observe them. Colonies are not at all consistent in their flowering habits from year to year and it often takes a great deal of persistence on the part of the observer to catch them in prime condition. *T. trianthophora* subsp. *mexicana* has occasionally been reported from Florida but no vouchers could be found. It is restricted to Mexico.

forma *caerulea*

forma *albidoflava*

Zeuxine

An African and Asiatic genus of about 30 species, *Zeuxine* is allied to *Spiranthes*, *Platythelys*, and *Goodyera*. One species, the lawn orchid, *Zeuxine strateumatica*, is naturalized in the New World. Plants appear to be both apomictic and annual. They often appear in greenhouses and potted nursery stock as well as throughout the landscape.

Zeuxine strateumatica (Linnaeus) Schlechter*

lawn orchid

Texas, east to Florida, and north to South Carolina; West Indies
Texas, Louisiana, Alabama, Mississippi, Georgia, Florida, South Carolina: well established only in Florida and perhaps introduced elsewhere
Plant: terrestrial, 4–25 cm tall
Leaves: 5–12; lanceolate, green with purple or tan pigmentation (the stem as well); 0.3–1.0 cm wide × 1–8 cm long
Flowers: 5–50+; in a densely flowered terminal spike, the flowers twisted to an angle; sepals and petals similar, oblanceolate; perianth white; lip narrowed at the base and broadly spreading at the apex, bright yellow; individual flower size 6–8 mm
Habitat: lawns, shrub borders, roadsides, and now established in out-of-the-way natural areas
Flowering period: (late October) December–March

This interesting introduction from Asia or Africa is becoming an abundant and attractive orchid throughout the Coastal Plain states. The lawn orchid often appears in unlikely spots around homes or even in nursery pots. The plants vary greatly in size and vigor, but no other species at that time of year can be confused with the distinctive white and yellow flowers. To some it may appear to be a *Spiranthes*, but closer examination will reveal a very different lip. The plants are annuals that move around as the seed blows. Capsules mature in a matter of one to two weeks after flowering. Despite its widespread habits it is in no danger of being a threat to our native orchid populations.

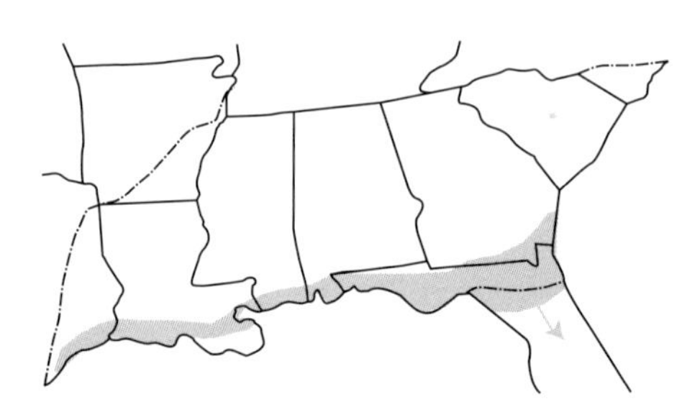

Three Species Found in Bordering Counties in North Carolina

All three of the following are found growing in the southern Appalachian Mountains in North Carolina in counties bordering on Georgia and/or South Carolina.

Coeloglossum viride (Linnaeus) Hartman var. *virescens* (Mühlenberg) Luer
long-bracted green orchis
Green-flowered plants are found in rich, often calcareous, woodlands, flowering in late spring (April–early May) in Jackson County, North Carolina.

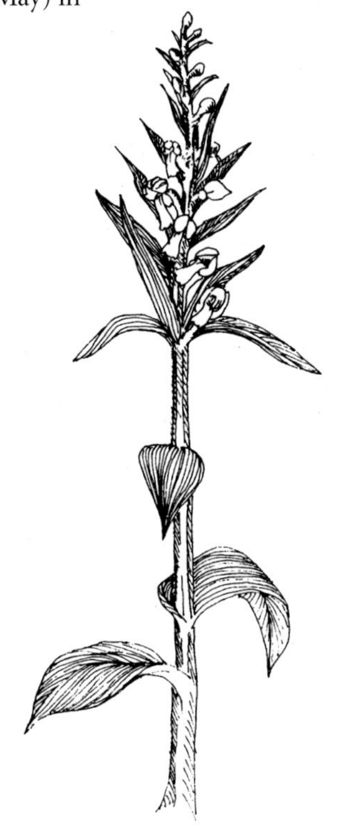

Cypripedium reginae Walter
showy lady's-slipper
The large, spectacular plants are found in calcareous bogs, seeps, and swamps. They flower in late May and early June in Macon and Jackson Counties, North Carolina.

Goodyera repens (Linnaeus) R. Brown forma *ophioides* (Fernald) P. M. Brown
lesser rattlesnake orchis
Plants of this diminutive gem are found in mountain woodlands of Macon County, North Carolina, and flower in late July.

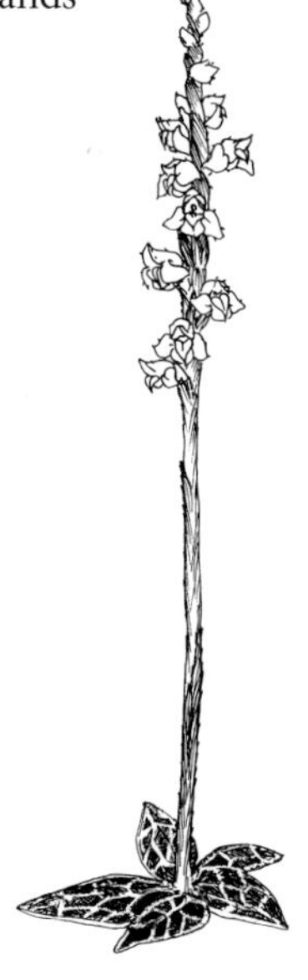

Note: The only species reported from the Southeast, as defined in this work, that has not been verified is *Platanthera orbiculata* (Pursh) Lindley, the pad-leaved orchis. It has been reported by Bentley in *Native Orchids of the Southern Appalachian Mountains* (p. 183) as occurring in northern Georgia, but no voucher can be found. It does occur northward in North Carolina and flowers in midsummer.

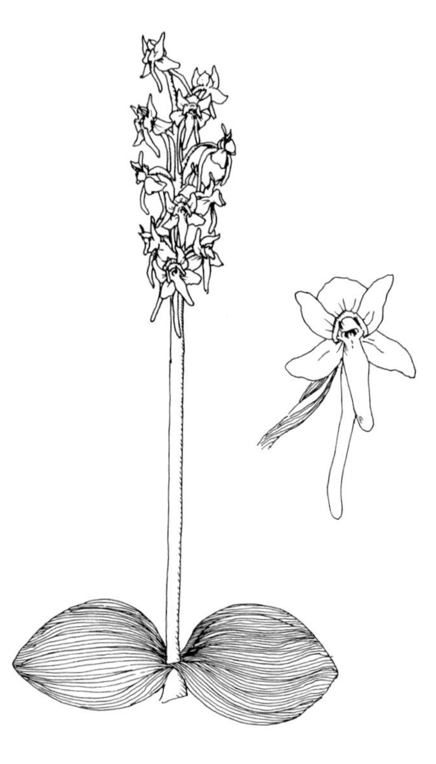

Part 3

References and Resources

Checklist of the Wild Orchids of the Southeastern United States, North of Peninsular Florida

Note: * = naturalized species

Aplectrum hyemale (Mühlenberg *ex* Willdenow) Nuttall
putty-root
forma *pallidum* House—yellow-flowered form

Arethusa bulbosa Linnaeus
dragon's-mouth
forma *albiflora* Rand & Redfield—white-flowered form
forma *subcaerulea* Rand & Redfield—lilac-blue-flowered form

Bletilla striata (Thunberg) Reichenbach *f.**
urn orchid

Calopogon barbatus (Walter) Ames
bearded grass-pink
forma *albiflorus* P. M. Brown—white-flowered form
forma *lilacinus* P. M. Brown—lilac-flowered form

Calopogon multiflorus Lindley
many-flowered grass-pink

Calopogon oklahomensis D. H. Goldman
Oklahoma grass-pink
forma *albiflorus* P. M. Brown—white-flowered form

Calopogon pallidus Chapman
pale grass-pink
forma *albiflorus* P. M. Brown—white-flowered form

Calopogon tuberosus (Linnaeus) Britton, Sterns, & Poggenberg var. *tuberosus*
common grass-pink
  forma *albiflorus* Britton—white-flowered form

Cleistes bifaria (Fernald) Catling & Gregg
upland spreading pogonia

Cleistes divaricata (Linnaeus) Ames
spreading pogonia
  forma *leucantha* P. M. Brown—white-flowered form

Corallorhiza maculata (Rafinesque) Rafinesque var. *maculata*
spotted coralroot
  forma *flavida* (Peck) Farwell—yellow-stemmed form
  forma *rubra* P. M. Brown—red-stemmed form

Corallorhiza odontorhiza (Willdenow) Poiret var. *odontorhiza*
autumn coralroot
  forma *flavida* Wherry—yellow-stemmed form
Corallorhiza odontorhiza (Willdenow) Poiret var. *pringlei* (Greenman) Freudenstein
Pringle's autumn coralroot

Corallorhiza wisteriana Conrad
Wister's coralroot
  forma *albolabia* P. M. Brown—white-flowered form
  forma *rubra* P. M. Brown—red-stemmed form

Cypripedium acaule Aiton
pink lady's-slipper, moccasin flower
  forma *albiflorum* Rand & Redfield—white-flowered form
  forma *biflorum* P. M. Brown—two-flowered form

Cypripedium candidum Mühlenberg *ex* Willdenow
small white lady's-slipper

Cypripedium kentuckiense C. F. Reed
ivory-lipped lady's-slipper
  forma *pricei* P. M. Brown—white-flowered form
  forma *summersii* P. M. Brown—concolorous yellow-flowered form

Cypripedium parviflorum Salisbury var. *parviflorum*
southern small yellow lady's-slipper
  forma *albolabium* Magrath & Norman—white-lipped form

Cypripedium parviflorum Salisbury var. *pubescens* (Willdenow) Knight
large yellow lady's-slipper

Epidendrum magnoliae Mühlenberg var. *magnoliae*
green-fly orchis

Epipactis helleborine (Linnaeus) Cranz*
broad-leaved helleborine
forma *alba* (Webster) Boivin—white-flowered form
forma *luteola* P. M. Brown—yellow-flowered form
forma *monotropoides* (Mousley) Scoggin—albino form
forma *variegata* (Webster) Boivin—variegated-leaved form
forma *viridens* A. Gray—green-flowered form

Eulophia alta (Linnaeus) Fawcett & Rendle
wild coco
forma *pallida* P. M. Brown—pale-flowered form
forma *pelchatii* P. M. Brown—white and green-flowered form

Galearis spectabilis (Linnaeus) Rafinesque
showy orchis
forma *gordinierii* (House) Whiting & Catling—white-flowered form
forma *willeyi* (Seymour) P. M. Brown—pink-flowered form

Goodyera pubescens (Willdenow) R. Brown
downy rattlesnake orchis

Gymnadeniopsis clavellata (Michaux) Rydberg var. *clavellata*
little club-spur orchis
forma *slaughteri* (P. M. Brown) P. M. Brown—white-flowered form

Gymnadeniopsis integra (Nuttall) Rydberg
yellow fringeless orchis

Gymnadeniopsis nivea (Nuttall) Rydberg
snowy orchis

Habenaria odontopetala Reichenbach *f.*
toothed rein orchis
forma *heatonii* P. M. Brown—albino form

Habenaria quinqueseta (Michaux) Eaton
Michaux's orchis

Habenaria repens Nuttall
water spider orchis

Hexalectris spicata (Walter) Barnhardt var. *spicata*
crested coralroot
forma *albolabia* P. M. Brown—white-flowered form

Isotria medeoloides (Pursh) Rafinesque
small whorled pogonia

Isotria verticillata (Mühlenberg *ex* Willdenow) Rafinesque
large whorled pogonia

Liparis liliifolia (Linnaeus) Richard *ex* Lindley
lily-leaved twayblade
forma *viridiflora* Wadmond—green-flowered form

Liparis loeselii (Linnaeus) Richard
Löesel's twayblade, fen orchis

Hybrid:
Liparis ×*jonesii* S. Bentley
Jones' hybrid twayblade
(*L. liliifolia* × *L. loeselii*)

Listera australis Lindley
southern twayblade
forma *scottii* P. M. Brown—many-leaved form
forma *trifolia* P. M. Brown—three-leaved form
forma *viridis* P. M. Brown—green-flowered form

Listera smallii Wiegand
Small's twayblade
forma *variegata* P. M. Brown—variegated-leaved form

Malaxis bayardii Fernald
Bayard's adder's-mouth

Malaxis spicata Swartz
Florida adder's-mouth
forma *trifoliata* P. M. Brown—three-leaved form

Malaxis unifolia Michaux
green adder's-mouth
forma *bifolia* Mousley—two-leaved form
forma *variegata* Mousley—variegated-leaf form

Platanthera blephariglottis (Willdenow) Lindley
northern white fringed orchis
forma *holopetala* (Lindley) P. M. Brown—entire-lip form

Platanthera chapmanii (Small) Luer *emend.* Folsom
Chapman's fringed orchis

Platanthera ciliaris (Linnaeus) Lindley
orange fringed orchis

Platanthera conspicua (Nash) P. M. Brown
southern white fringed orchis

Platanthera cristata (Michaux) Lindley
orange crested orchis
forma *straminea* P. M. Brown—pale yellow-flowered form

Platanthera flava (Linnaeus) Lindley var. *flava*
southern tubercled orchis
Platanthera flava (Linnaeus) Lindley var. *herbiola* (R. Brown) Luer
northern tubercled orchis
forma *lutea* (Boivin) Whiting & Catling—yellow-flowered form

Platanthera grandiflora (Bigelow) Lindley
large purple fringed orchis
forma *albiflora* (Rand & Redfield) Catling—white-flowered form
forma *bicolor* P. M. Brown—bicolor-flowered form
forma *carnea* P. M. Brown—pink-flowered form
forma *mentotonsa* (Fernald) P. M. Brown—entire-lip form

Platanthera integrilabia (Correll) Luer
monkey-face orchis

Platanthera lacera (Michaux) G. Don
green fringed orchis, ragged orchis

Platanthera leucophaea (Nuttall) Lindley
eastern prairie fringed orchis

Platanthera peramoena (A. Gray) A. Gray
purple fringeless orchis
forma *doddiae* P. M. Brown—white-flowered form

Platanthera psycodes (Linnaeus) Lindley
small purple fringed orchis
forma *albiflora* (R. Hoffman) Whiting & Catling—white-flowered form
forma *ecalcarata* (Bryan) P. M. Brown—spurless form
forma *rosea* P. M. Brown—pink-flowered form
forma *varians* (Bryan) P. M. Brown—entire-lip form

Hybrids:

Platanthera ×*andrewsii* (Niles) Luer
Andrews' hybrid fringed orchis
(*P. lacera* × *P. psycodes*)

Platanthera ×*apalachicola* P. M. Brown & S. Stewart
Apalachicola hybrid fringed orchis
(*P. chapmanii* × *P. cristata*)

Platanthera ×*beckneri* P. M. Brown
Beckner's hybrid fringed orchis
(*P. conspicua* × *P. cristata*)

Platanthera ×*bicolor* (Rafinesque) Luer
bicolor hybrid fringed orchis
(*P. blephariglottis* × *P. ciliaris*)

Platanthera ×*canbyi* (Ames) Luer
Canby's hybrid fringed orchis
(*P. blephariglottis* × *P. cristata*)

Platanthera ×*channellii* Folsom
Channell's hybrid fringed orchis
(*P. ciliaris* × *P. cristata*)

Platanthera ×*keenanii* P. M. Brown
Keenan's hybrid fringed orchis
(*P. grandiflora* × *P. lacera*)

Platanthera ×*lueri* P. M. Brown
Luer's hybrid fringed orchis
(*P. ciliaris* × *P. conspicua*)

Platanthera ×*osceola* P. M. Brown & S. Stewart
Osceola hybrid fringed orchis
(*P. chapmanii* × *P. ciliaris*)

Platythelys querceticola (Lindley) Garay
low ground orchid, jug orchid

Pogonia ophioglossoides (Linnaeus) Ker-Gawler
rose pogonia
  forma *albiflora* Rand & Redfield—white-flowered form
  forma *brachypogon* (Fernald) P. M. Brown—short-bearded form

Ponthieva racemosa (Walter) C. Mohr
shadow-witch

Pteroglossaspis ecristata (Fernald) Rolfe
crestless plume orchid
  forma *flava* P. M. Brown—yellow-flowered form
  forma *purpurea* P. M. Brown—dusky purple-flowered form

Sacoila lanceolata (Aublet) Garay var. *lanceolata*
leafless beaked orchid
  forma *albidaviridis* Catling & Sheviak—white and green-flowered form
  forma *folsomii* P. M. Brown—golden bronze-flowered form

Spiranthes brevilabris Lindley
short-lipped ladies'-tresses, Texas ladies'-tresses

Spiranthes cernua (Linnaeus) L. C. Richard
nodding ladies'-tresses

Spiranthes eatonii Ames *ex* P. M. Brown
Eaton's ladies'-tresses

Spiranthes floridana (Wherry) Cory *emend.* P. M. Brown
Florida ladies'-tresses

Spiranthes lacera (Rafinesque) Rafinesque var. *gracilis* (Bigelow) Luer
southern slender ladies'-tresses

Spiranthes laciniata (Small) Ames
lace-lipped ladies'-tresses

Spiranthes longilabris Lindley
long-lipped ladies'-tresses

Spiranthes lucida (H. H. Eaton) Ames
shining ladies'-tresses

Spiranthes magnicamporum Sheviak
Great Plains ladies'-tresses

Spiranthes ochroleuca (Rydberg) Rydberg
yellow ladies'-tresses

Spiranthes odorata (Nuttall) Lindley
fragrant ladies'-tresses

Spiranthes ovalis Lindley var. *ovalis*
southern oval ladies'-tresses
Spiranthes ovalis Lindley var. *erostellata* Catling
northern oval ladies'-tresses

Spiranthes parksii Correll
Navasota ladies'-tresses

Spiranthes praecox (Walter) S. Watson
giant ladies'-tresses
forma *albolabia* P. M. Brown & C. McCartney—white-lipped form

Spiranthes sylvatica P. M. Brown
woodland ladies'-tresses

Spiranthes tuberosa Rafinesque
little ladies'-tresses

Spiranthes vernalis Engelmann & Gray
grass-leaved ladies'-tresses

Hybrids:

Spiranthes ×folsomii P. M. Brown
Folsom's hybrid ladies'-tresses
(*S. longilabris* × *S. odorata*)

Spiranthes ×itchetuckneensis P. M. Brown
Ichetucknee Springs hybrid ladies'-tresses
(*S. odorata* × *S. ovalis* var. *ovalis*)

Spiranthes ×intermedia Ames
intermediate hybrid ladies'-tresses
(*S. lacera* var. *gracilis* × *S. vernalis*)

Spiranthes ×meridionalis P. M. Brown
southern hybrid ladies'-tresses
(*S. praecox* × *S. vernalis*)

Tipularia discolor (Pursh) Nuttall
crane-fly orchis
forma *viridifolia* P. M. Brown—green-leaved form

Triphora rickettii Luer
Rickett's noddingcaps

Triphora trianthophora (Swartz) Rydberg subsp. *trianthophora*
three birds orchid
forma *albidoflava* Keenan—white-flowered form
forma *caerulea* P. M. Brown—blue-flowered form
forma *rossii* P. M. Brown—multi-colored form

Zeuxine strateumatica (Linnaeus) Schlechter*
lawn orchid

State Checklists of Orchids Found in the Southeastern United States

Note: [] = unvouchered reports, * = naturalized species

Alabama

58 species and varieties, including 2 unvouchered reports

Aplectrum hyemale, putty-root
Calopogon barbatus, bearded grass-pink
Calopogon multiflorus, many-flowered grass-pink
Calopogon oklahomensis, Oklahoma grass-pink
Calopogon pallidus, pale grass-pink
Calopogon tuberosus, common grass-pink
Cleistes bifaria, upland spreading pogonia
Corallorhiza odontorhiza var. *odontorhiza,* autumn coralroot
Corallorhiza wisteriana, Wister's coralroot
Cypripedium acaule, pink lady's-slipper, moccasin flower
Cypripedium candidum, small white lady's-slipper
Cypripedium kentuckiense, ivory-lipped lady's-slipper
Cypripedium parviflorum var. *parviflorum,* southern small yellow lady's-slipper
[*Cypripedium parviflorum* var. *pubescens,* large yellow lady's-slipper]
Epidendrum magnoliae, green-fly orchis
Galearis spectabilis, showy orchis
Goodyera pubescens, downy rattlesnake orchis
Gymnadeniopsis clavellata, little club-spur orchis
Gymnadeniopsis integra, yellow fringeless orchis
Gymnadeniopsis nivea, snowy orchis
Habenaria quinqueseta, Michaux's orchid
Habenaria repens, water spider orchid
Hexalectris spicata, crested coralroot

Isotria verticillata, large whorled pogonia
Liparis liliifolia, lily-leaved twayblade
Liparis loeselii, Löesel's twayblade, fen orchis
Listera australis, southern twayblade
Malaxis unifolia, green adder's-mouth
Platanthera ciliaris, orange fringed orchis
Platanthera conspicua, southern white fringed orchis
Platanthera cristata, orange crested orchis
Platanthera flava var. *flava,* southern tubercled orchis
Platanthera integrilabia, monkey-face orchis
Platanthera lacera, green fringed orchis, ragged orchis
Platanthera peramoena, purple fringeless orchis
[*Platythelys querceticola,* low ground orchid, jug orchid]
Pogonia ophioglossoides, rose pogonia
Ponthieva racemosa, shadow-witch
Pteroglossaspis ecristata, crestless plume orchid
Spiranthes brevilabris, short-lipped ladies'-tresses
Spiranthes cernua, nodding ladies'-tresses
Spiranthes eatonii, Eaton's ladies'-tresses
Spiranthes floridana, Florida ladies'-tresses
Spiranthes lacera var. *gracilis,* southern slender ladies'-tresses
Spiranthes laciniata, lace-lipped ladies'-tresses
Spiranthes longilabris, long-lipped ladies'-tresses
Spiranthes lucida, shining ladies'-tresses
Spiranthes magnicamporum, Great Plains ladies'-tresses
Spiranthes odorata, fragrant ladies'-tresses
Spiranthes ovalis var. *ovalis,* southern oval ladies'-tresses
Spiranthes ovalis var. *erostellata,* northern oval ladies'-tresses
Spiranthes praecox, giant ladies'-tresses
Spiranthes sylvatica, woodland ladies'-tresses
Spiranthes tuberosa, little ladies'-tresses
Spiranthes vernalis, grass-leaved ladies'-tresses
Tipularia discolor, crane-fly orchis
Triphora trianthophora, three birds orchid
Zeuxine strateumatica, lawn orchid*

Arkansas

28 species and varieties found in the Coastal Plain and Prairie regions, of 37 species found in the state

Calopogon oklahomensis, Oklahoma grass-pink
Calopogon tuberosus, common grass-pink
Cypripedium kentuckiense, ivory-lipped lady's-slipper
Galearis spectabilis, showy orchis
Gymnadeniopsis clavellata, little club-spur orchis
Gymnadeniopsis nivea, snowy orchis
Habenaria repens, water spider orchid
Hexalectris spicata, crested coralroot
Isotria verticillata, large whorled pogonia
Listera australis, southern twayblade
Malaxis unifolia, green adder's-mouth
Platanthera ciliaris, orange fringed orchis
Platanthera cristata, orange crested orchis
Platanthera flava var. *flava,* southern tubercled orchis
Platanthera lacera, green fringed orchis, ragged orchis
Pogonia ophioglossoides, rose pogonia
Ponthieva racemosa, shadow-witch
Spiranthes cernua, nodding ladies'-tresses
Spiranthes lacera var. *gracilis,* southern slender ladies'-tresses
Spiranthes magnicamporum, Great Plains ladies'-tresses
Spiranthes odorata, fragrant ladies'-tresses
Spiranthes ovalis var. *ovalis,* southern oval ladies'-tresses
Spiranthes ovalis var. *erostellata,* northern oval ladies'-tresses
Spiranthes sylvatica, woodland ladies'-tresses (*S. praecox*?)
Spiranthes tuberosa, little ladies'-tresses
Spiranthes vernalis, grass-leaved ladies'-tresses
Tipularia discolor, crane-fly orchis
Triphora trianthophora, three birds orchid

Florida

51 species and varieties found north of peninsular Florida, of 118 species and varieties found within the state

Bletilla striata, urn orchid*
Calopogon barbatus, bearded grass-pink
Calopogon multiflorus, many-flowered grass-pink
Calopogon pallidus, pale grass-pink
Calopogon tuberosus, common grass-pink
Cleistes bifaria, upland spreading pogonia
Cleistes divaricata, spreading pogonia

Corallorhiza odontorhiza var. *odontorhiza*, autumn coralroot
Corallorhiza wisteriana, Wister's coralroot
Epidendrum magnoliae var. *magnoliae*, green-fly orchis
Eulophia alta, wild coco
Goodyera pubescens, downy rattlesnake orchis
Gymnadeniopsis clavellata, little club-spur orchis
Gymnadeniopsis integra, yellow fringeless orchis
Gymnadeniopsis nivea, snowy orchis
Habenaria odontopetala, toothed rein orchis
Habenaria quinqueseta, Michaux's orchid
Habenaria repens, water spider orchid
Hexalectris spicata, crested coralroot
Isotria verticillata, large whorled pogonia
Listera australis, southern twayblade
Malaxis spicata, Florida adder's-mouth
Malaxis unifolia, green adder's-mouth
Mesadenus lucayanus, copper ladies'-tresses
Platanthera chapmanii, Chapman's fringed orchis
Platanthera ciliaris, orange fringed orchis
Platanthera conspicua, southern white fringed orchis
Platanthera cristata, orange crested orchis
Platanthera flava var. *flava*, southern tubercled orchis
Platythelys querceticola, low ground orchid, jug orchid
Pogonia ophioglossoides, rose pogonia
Ponthieva racemosa, shadow-witch
Pteroglossaspis ecristata, crestless plume orchid
Sacoila lanceolata, leafless beaked orchid
Spiranthes brevilabris, short-lipped ladies'-tresses, Texas ladies'-tresses
Spiranthes cernua, nodding ladies'-tresses
Spiranthes eatonii, Eaton's ladies'-tresses
Spiranthes floridana, Florida ladies'-tresses
Spiranthes laciniata, lace-lipped ladies'-tresses
Spiranthes longilabris, long-lipped ladies'-tresses
Spiranthes odorata, fragrant ladies'-tresses
Spiranthes ovalis var. *ovalis*, southern oval ladies'-tresses
Spiranthes ovalis var. *erostellata*, northern oval ladies'-tresses
Spiranthes praecox, giant ladies'-tresses
Spiranthes sylvatica, woodland ladies'-tresses
Spiranthes tuberosa, little ladies'-tresses
Spiranthes vernalis, grass-leaved ladies'-tresses

Tipularia discolor, crane-fly orchis
Triphora rickettii, Rickett's noddingcaps
Triphora trianthophora, three birds orchid
Zeuxine strateumatica, lawn orchid*

Georgia

67 species and varieties, including one unvouchered report

Aplectrum hyemale, putty-root
Calopogon barbatus, bearded grass-pink
Calopogon multiflorus, many-flowered grass-pink
Calopogon oklahomensis, Oklahoma grass-pink
Calopogon pallidus, pale grass-pink
Calopogon tuberosus, common grass-pink
Cleistes bifaria, upland spreading pogonia
Cleistes divaricata, spreading pogonia
Corallorhiza maculata, spotted coralroot
Corallorhiza odontorhiza var. *odontorhiza*, autumn coralroot
Corallorhiza odontorhiza var. *pringlei*, Pringle's autumn coralroot
Corallorhiza wisteriana, Wister's coralroot
Cypripedium acaule, pink lady's-slipper, moccasin flower
Cypripedium kentuckiense, ivory-lipped lady's-slipper
Cypripedium parviflorum var. *parviflorum*, southern small yellow lady's-slipper
Cypripedium parviflorum var. *pubescens*, large yellow lady's-slipper
Epidendrum magnoliae, green-fly orchis
Epipactis helleborine, broad-leaved helleborine*
Eulophia alta, wild coco
Galearis spectabilis, showy orchis
Goodyera pubescens, downy rattlesnake orchis
Gymnadeniopsis clavellata, little club-spur orchis
Gymnadeniopsis integra, yellow fringeless orchis
Gymnadeniopsis nivea, snowy orchis
Habenaria quinqueseta, Michaux's orchid
Habenaria repens, water spider orchid
Hexalectris spicata, crested coralroot
Isotria medeoloides, small whorled pogonia
Isotria verticillata, large whorled pogonia
Liparis liliifolia, lily-leaved twayblade
Listera australis, southern twayblade
Listera smallii, Small's twayblade

Malaxis spicata, Florida adder's-mouth
Malaxis unifolia, green adder's-mouth
Platanthera blephariglottis, northern white fringed orchis
Platanthera chapmanii, Chapman's fringed orchis
Platanthera ciliaris, orange fringed orchis
Platanthera conspicua, southern white fringed orchis
Platanthera cristata, orange crested orchis
Platanthera flava var. *flava,* southern tubercled orchis
Platanthera flava var. *herbiola,* northern tubercled orchis
Platanthera grandiflora, large purple fringed orchis
Platanthera integrilabia, monkey-face orchis
Platanthera lacera, green fringed orchis, ragged orchis
[*Platanthera orbiculata,* pad-leaved orchis]
Platanthera peramoena, purple fringeless orchis
Platanthera psycodes, small purple fringed orchis
Pogonia ophioglossoides, rose pogonia
Ponthieva racemosa, shadow-witch
Pteroglossaspis ecristata, crestless plume orchid
Spiranthes cernua, nodding ladies'-tresses
Spiranthes eatonii, Eaton's ladies'-tresses
Spiranthes floridana, Florida ladies'-tresses
Spiranthes lacera var. *gracilis,* southern slender ladies'-tresses
Spiranthes laciniata, lace-lipped ladies'-tresses
Spiranthes longilabris, long-lipped ladies'-tresses
Spiranthes magnicamporum, Great Plains ladies'-tresses
Spiranthes odorata, fragrant ladies'-tresses
Spiranthes ovalis var. *ovalis,* southern oval ladies'-tresses
Spiranthes ovalis var. *erostellata,* northern oval ladies'-tresses
Spiranthes praecox, giant ladies'-tresses
Spiranthes sylvatica, woodland ladies'-tresses
Spiranthes tuberosa, little ladies'-tresses
Spiranthes vernalis, grass-leaved ladies'-tresses
Tipularia discolor, crane-fly orchis
Triphora trianthophora, three birds orchid
Zeuxine strateumatica, lawn orchid*

Louisiana

46 species and varieties
Calopogon barbatus, bearded grass-pink

Calopogon multiflorus, many-flowered grass-pink
Calopogon oklahomensis, Oklahoma grass-pink
Calopogon pallidus, pale grass-pink
Calopogon tuberosus, common grass-pink
Corallorhiza odontorhiza var. *odontorhiza*, autumn coralroot
Corallorhiza wisteriana, Wister's coralroot
Cypripedium kentuckiense, ivory-lipped lady's-slipper
Epidendrum magnoliae, green-fly orchis
Gymnadeniopsis clavellata, little club-spur orchis
Gymnadeniopsis integra, yellow fringeless orchis
Gymnadeniopsis nivea, snowy orchis
Habenaria quinqueseta, Michaux's orchid
Habenaria repens, water spider orchid
Hexalectris spicata, crested coralroot
Isotria verticillata, large whorled pogonia
Listera australis, southern twayblade
Malaxis unifolia, green adder's-mouth
Platanthera ciliaris, orange fringed orchis
Platanthera conspicua, southern white fringed orchis
Platanthera cristata, orange crested orchis
Platanthera flava var. *flava*, southern tubercled orchis
Platanthera lacera, green fringed orchis, ragged orchis
Platanthera leucophaea, eastern prairie fringed orchis
Platythelys querceticola, low ground orchid, jug orchid
Pogonia ophioglossoides, rose pogonia
Ponthieva racemosa, shadow-witch
Pteroglossaspis ecristata, crestless plume orchid
Spiranthes brevilabris, short-lipped ladies'-tresses, Texas ladies'-tresses
Spiranthes cernua, nodding ladies'-tresses
Spiranthes eatonii, Eaton's ladies'-tresses
Spiranthes floridana, Florida ladies'-tresses
Spiranthes lacera var. *gracilis*, southern slender ladies'-tresses
Spiranthes laciniata, lace-lipped ladies'-tresses
Spiranthes longilabris, long-lipped ladies'-tresses
Spiranthes magnicamporum, Great Plains ladies'-tresses
Spiranthes odorata, fragrant ladies'-tresses
Spiranthes ovalis var. *ovalis*, southern oval ladies'-tresses
Spiranthes ovalis var. *erostellata*, northern oval ladies'-tresses
Spiranthes praecox, giant ladies'-tresses
Spiranthes sylvatica, woodland ladies'-tresses

Spiranthes tuberosa, little ladies'-tresses
Spiranthes vernalis, grass-leaved ladies'-tresses
Tipularia discolor, crane-fly orchis
Triphora trianthophora, three birds orchid
Zeuxine strateumatica, lawn orchid*

Mississippi

54 species and varieties, including 1 unvouchered report

Calopogon barbatus, bearded grass-pink
Calopogon multiflorus, many-flowered grass-pink
Calopogon oklahomensis, Oklahoma grass-pink
Calopogon pallidus, pale grass-pink
Calopogon tuberosus, common grass-pink
Cleistes bifaria, upland spreading pogonia
Corallorhiza odontorhiza var. *odontorhiza,* autumn coralroot
Corallorhiza wisteriana, Wister's coralroot
Cypripedium acaule, pink lady's-slipper, moccasin flower
Cypripedium kentuckiense, ivory-lipped lady's-slipper
Cypripedium parviflorum var. *pubescens,* large yellow lady's-slipper
Epidendrum magnoliae, green-fly orchis
Galearis spectabilis, showy orchis
Goodyera pubescens, downy rattlesnake orchis
Gymnadeniopsis clavellata, little club-spur orchis
Gymnadeniopsis integra, yellow fringeless orchis
Gymnadeniopsis nivea, snowy orchis
Habenaria quinqueseta, Michaux's orchid
Habenaria repens, water spider orchid
Hexalectris spicata, crested coralroot
Isotria verticillata, large whorled pogonia
Liparis liliifolia, lily-leaved twayblade
Liparis loeselii, Löesel's twayblade, fen orchis
Listera australis, southern twayblade
Malaxis unifolia, green adder's-mouth
Platanthera ciliaris, orange fringed orchis
Platanthera conspicua, southern white fringed orchis
Platanthera cristata, orange crested orchis
Platanthera flava var. *flava,* southern tubercled orchis
Platanthera integrilabia, monkey-face orchis
Platanthera lacera, green fringed orchis, ragged orchis

Platanthera peramoena, purple fringeless orchis
[*Platythelys querceticola,* low ground orchid, jug orchid]
Pogonia ophioglossoides, rose pogonia
Ponthieva racemosa, shadow-witch
Pteroglossaspis ecristata, crestless plume orchid
Spiranthes brevilabris, short-lipped ladies'-tresses
Spiranthes cernua, nodding ladies'-tresses
Spiranthes eatonii, Eaton's ladies'-tresses
Spiranthes floridana, Florida ladies'-tresses
Spiranthes lacera var. *gracilis,* southern slender ladies'-tresses
Spiranthes laciniata, lace-lipped ladies'-tresses
Spiranthes longilabris, long-lipped ladies'-tresses
Spiranthes magnicamporum, Great Plains ladies'-tresses
Spiranthes odorata, fragrant ladies'-tresses
Spiranthes ovalis var. *ovalis,* southern oval ladies'-tresses
Spiranthes ovalis var. *erostellata,* northern oval ladies'-tresses
Spiranthes praecox, giant ladies'-tresses
Spiranthes sylvatica, woodland ladies'-tresses
Spiranthes tuberosa, little ladies'-tresses
Spiranthes vernalis, grass-leaved ladies'-tresses
Tipularia discolor, crane-fly orchis
Triphora trianthophora, three birds orchid
Zeuxine strateumatica, lawn orchid*

North Carolina

43 species found in the southeastern counties of the Coastal Plain, of 48 species and varieties found within the state

Calopogon barbatus, bearded grass-pink
Calopogon multiflorus, many-flowered grass-pink
Calopogon pallidus, pale grass-pink
Calopogon tuberosus, common grass-pink
Cleistes bifaria, upland spreading pogonia
Cleistes divaricata, spreading pogonia
Corallorhiza odontorhiza var. *odontorhiza,* autumn coralroot
Corallorhiza wisteriana, Wister's coralroot
Cypripedium acaule, pink lady's-slipper, moccasin flower
Epidendrum magnoliae, green-fly orchis
Goodyera pubescens, downy rattlesnake orchis
Gymnadeniopsis clavellata, little club-spur orchis

Gymnadeniopsis integra, yellow fringeless orchis
Gymnadeniopsis nivea, snowy orchis
Habenaria quinqueseta, Michaux's orchid
Habenaria repens, water spider orchid
Hexalectris spicata, crested coralroot
Isotria verticillata, large whorled pogonia
Listera australis, southern twayblade
Malaxis spicata, Florida adder's-mouth
Malaxis unifolia, green adder's-mouth
Platanthera blephariglottis, northern white fringed orchis
Platanthera ciliaris, orange fringed orchis
Platanthera conspicua, southern white fringed orchis
Platanthera cristata, orange crested orchis
Platanthera flava var. *flava,* southern tubercled orchis
Pogonia ophioglossoides, rose pogonia
Ponthieva racemosa, shadow-witch
Pteroglossaspis ecristata, crestless plume orchid
Spiranthes cernua, nodding ladies'-tresses
Spiranthes eatonii, Eaton's ladies'-tresses
Spiranthes floridana, Florida ladies'-tresses
Spiranthes lacera var. *gracilis,* southern slender ladies'-tresses
Spiranthes laciniata, lace-lipped ladies'-tresses
Spiranthes longilabris, long-lipped ladies'-tresses
Spiranthes odorata, fragrant ladies'-tresses
Spiranthes ovalis var. *erostellata,* northern oval ladies'-tresses
Spiranthes praecox, giant ladies'-tresses
Spiranthes sylvatica, woodland ladies'-tresses
Spiranthes tuberosa, little ladies'-tresses
Spiranthes vernalis, grass-leaved ladies'-tresses
Tipularia discolor, crane-fly orchis
Triphora trianthophora, three birds orchid

South Carolina

60 species and varieties, including one unvouchered report

Aplectrum hyemale, putty-root, Adam and Eve
Arethusa bulbosa, dragon's-mouth
Calopogon barbatus, bearded grass-pink
Calopogon multiflorus, many-flowered grass-pink
Calopogon oklahomensis, Oklahoma grass-pink

Calopogon pallidus, pale grass-pink
Calopogon tuberosus, common grass-pink
Cleistes bifaria, upland spreading pogonia
Cleistes divaricata, spreading pogonia
Corallorhiza maculata, spotted coralroot
Corallorhiza odontorhiza var. *odontorhiza,* autumn coralroot
Corallorhiza wisteriana, Wister's coralroot
Cypripedium acaule, pink lady's-slipper, moccasin flower
Cypripedium parviflorum var. *pubescens,* large yellow lady's-slipper
Epidendrum magnoliae, green-fly orchis
Galearis spectabilis, showy orchis
Goodyera pubescens, downy rattlesnake orchis
Gymnadeniopsis clavellata, little club-spur orchis
Gymnadeniopsis integra, yellow fringeless orchis
Gymnadeniopsis nivea, snowy orchis
Habenaria quinqueseta, Michaux's orchid
Habenaria repens, water spider orchid
Hexalectris spicata, crested coralroot
Isotria medeoloides, small whorled pogonia
Isotria verticillata, large whorled pogonia
Liparis liliifolia, lily-leaved twayblade
Listera australis, southern twayblade
Listera smallii, Small's twayblade
Malaxis bayardii, Bayard's adder's-mouth
Malaxis spicata, Florida adder's-mouth
Malaxis unifolia, green adder's-mouth
Platanthera blephariglottis, northern white fringed orchis
Platanthera ciliaris, orange fringed orchis
Platanthera conspicua, southern white fringed orchis
Platanthera cristata, orange crested orchis
Platanthera flava var. *flava,* southern tubercled orchis
[*Platanthera flava* var. *herbiola,* northern tubercled orchis]
Platanthera integrilabia, monkey-face orchis
Platanthera lacera, green fringed orchis, ragged orchis
Platanthera peramoena, purple fringeless orchis
Platanthera psycodes, small purple fringed orchis
Pogonia ophioglossoides, rose pogonia
Ponthieva racemosa, shadow-witch
Pteroglossaspis ecristata, crestless plume orchid
Spiranthes cernua, nodding ladies'-tresses

Spiranthes eatonii, Eaton's ladies'-tresses
Spiranthes floridana, Florida ladies'-tresses
Spiranthes lacera var. *gracilis*, southern slender ladies'-tresses
Spiranthes laciniata, lace-lipped ladies'-tresses
Spiranthes longilabris, long-lipped ladies'-tresses
Spiranthes ochroleuca, yellow ladies'-tresses
Spiranthes odorata, fragrant ladies'-tresses
Spiranthes ovalis var. *erostellata*, northern oval ladies'-tresses
Spiranthes praecox, giant ladies'-tresses
Spiranthes sylvatica, woodland ladies'-tresses
Spiranthes tuberosa, little ladies'-tresses
Spiranthes vernalis, grass-leaved ladies'-tresses
Tipularia discolor, crane-fly orchis
Triphora trianthophora, three birds orchid
Zeuxine strateumatica, lawn orchid*

Texas

41 species and varieties found within the Piney Woods and Coastal Plain regions, of 54 species found within the state

Calopogon oklahomensis, Oklahoma grass-pink
Calopogon tuberosus, common grass-pink
Cleistes bifaria, upland spreading pogonia
Corallorhiza odontorhiza var. *odontorhiza*, autumn coralroot
Corallorhiza wisteriana, Wister's coralroot
Cypripedium kentuckiense, ivory-lipped lady's-slipper
Gymnadeniopsis clavellata, little club-spur orchis
Gymnadeniopsis integra, yellow fringeless orchis
Gymnadeniopsis nivea, snowy orchis
Habenaria quinqueseta, Michaux's orchid
Habenaria repens, waterspider orchis
Hexalectris spicata, crested coralroot
Isotria verticillata, large whorled pogonia
Listera australis, southern twayblade
Malaxis unifolia, green adder's-mouth
Platanthera chapmanii, Chapman's fringed orchis
Platanthera ciliaris, orange fringed orchis
Platanthera conspicua, southern white fringed orchis
Platanthera cristata, orange crested orchis
Platanthera flava var. *flava*, southern tubercled orchis

Platanthera lacera, green fringed orchis, ragged orchis
Pogonia ophioglossoides, rose pogonia
Ponthieva racemosa, shadow-witch
Spiranthes brevilabris, short-lipped ladies'-tresses, Texas ladies'-tresses
Spiranthes cernua, nodding ladies'-tresses
Spiranthes eatonii, Eaton's ladies'-tresses
Spiranthes floridana, Florida ladies'-tresses
Spiranthes lacera var. *gracilis,* southern slender ladies'-tresses
Spiranthes laciniata, lace-lipped ladies'-tresses
Spiranthes longilabris, long-lipped ladies'-tresses
Spiranthes odorata, fragrant ladies'-tresses
Spiranthes ovalis var. *ovalis,* southern oval ladies'-tresses
Spiranthes ovalis var. *erostellata,* northern oval ladies'-tresses
Spiranthes parksii, Navasota ladies'-tresses
Spiranthes praecox, giant ladies'-tresses
Spiranthes sylvatica, woodland ladies'-tresses
Spiranthes tuberosa, little ladies'-tresses
Spiranthes vernalis, grass-leaved ladies'-tresses
Tipularia discolor, crane-fly orchis
Triphora trianthophora, three birds orchid
Zeuxine strateumatica, lawn orchid*

Some Regional Orchid Statistics

Facts and Figures about Our Southeastern Orchids

Of the 81 species and varieties treated within the southeastern United States the following summarizes the distribution:

Only 18 species of orchids have been found within every state in the entire region

Calopogon tuberosus, common grass-pink
Gymnadeniopsis clavellata, little club-spur orchis
Gymnadeniopsis nivea, snowy orchis
Hexalectris spicata, crested coralroot
Isotria verticillata, large whorled pogonia
Listera australis, southern twayblade
Malaxis unifolia, green adder's-mouth
Platanthera ciliaris, orange fringed orchis
Platanthera cristata, orange crested orchis
Platanthera flava var. *flava,* southern tubercled orchis
Pogonia ophioglossoides, rose pogonia
Ponthieva racemosa, shadow-witch
Spiranthes odorata, fragrant ladies'-tresses
Spiranthes praecox, giant ladies'-tresses (may not be present in Arkansas)
Spiranthes sylvatica, woodland ladies'-tresses
Spiranthes tuberosa, little ladies'-tresses
Spiranthes vernalis, grass-leaved ladies'-tresses
Triphora trianthophora, three birds orchid

Within the southeastern United States:

Two species of orchids have been found only in Alabama:

Cypripedium candidum, small white lady's-slipper
Spiranthes lucida, shining ladies'-tresses

Five species have been found only in Florida:

Bletilla striata, urn orchid*
Habenaria odontopetala, toothed rein orchis
Mesadenus lucayanus, copper ladies'-tresses
Sacoila lanceolata, leafless beaked orchid
Triphora rickettii, Rickett's noddingcaps

Four taxa of orchids have been found only in Georgia:

Corallorhiza odontorhiza var. *pringlei,* Pringle's autumn coralroot
Epipactis helleborine, broad-leaved helleborine*
Platanthera flava var. *herbiola,* northern tubercled orchis
Platanthera grandiflora, large purple fringed orchis

One species of orchid has been found only in Louisiana

Platanthera leucophaea, eastern prairie fringed orchis

Three species of orchids have been found only in South Carolina:

Arethusa bulbosa, dragon's-mouth
Malaxis bayardii, Bayard's adder's-mouth
Spiranthes ochroleuca, yellow ladies'-tresses

One species of orchid has been found only in Texas:

Spiranthes parksii, Navasota ladies'-tresses

Seven species of orchids have been found only in two states:

Corallorhiza maculata, spotted coralroot—Georgia, South Carolina
Cypripedium parviflorum var. *parviflorum,* southern small yellow lady's-slipper—Alabama, Georgia
Eulophia alta, wild coco—Florida, Georgia
Isotria medeoloides, small whorled pogonia—South Carolina, Georgia
Liparis loeselii, Löesel's twayblade, fen orchis—Mississippi, Alabama
Listera smallii, Small's twayblade—Georgia, South Carolina
Platanthera psycodes, small purple fringed orchis—Georgia, South Carolina

Six taxa of orchids have been found only in three states:

Aplectrum hyemale, putty-root—Alabama, Georgia, South Carolina
Cypripedium parviflorum var. *pubescens,* large yellow lady's-slipper—Georgia, Mississippi, South Carolina
Platanthera blephariglottis, northern white fringed orchis—Georgia, South Carolina, North Carolina
Platanthera chapmanii, Chapman's fringed orchis—Florida, Georgia, Texas
Platythelys querceticola, low ground orchid, jug orchid—Florida, Louisiana, Mississippi

Spiranthes brevilabris, short-lipped ladies'-tresses, Texas ladies'-tresses—Texas, Mississippi, Florida

Three species of orchids have been found only in four states:

Cleistes divaricata, spreading pogonia—Florida, Georgia, North Carolina, South Carolina

Liparis liliifolia, lily-leaved twayblade—Alabama, Georgia, Mississippi, South Carolina

Malaxis spicata, Florida adder's-mouth—Florida, Georgia, North Carolina, South Carolina

Only one species of orchid is considered extirpated from the region:

Platanthera leucophaea, eastern prairie fringed orchis—Louisiana

Only one species of orchid is an unvouchered report from the region:

Platanthera orbiculata, pad-leaved orchis—Georgia

Three species of orchids are non-natives:

Bletilla striata, urn orchid, possibly a garden escape and known from a single collection in Escambia County, Florida

Epipactis helleborine, broad-leaved helleborine, the most widespread non-native that is thoroughly naturalized in much of North America, but only recently in northern Georgia

Zeuxine strateumatica, lawn orchid, thoroughly naturalized throughout Florida and common in nursery stock and greenhouses, most probably introduced to the other southeastern states via potted nursery stock

Only one species of orchid is truly endemic to the southeastern United States:

Platanthera chapmanii, Chapman's fringed orchis

Several other species of orchids have their primary distribution in the southeastern United States but they range either north in the mountains or along the Coastal Plain:

Calopogon barbatus, bearded grass-pink
Calopogon multiflorus, many-flowered grass-pink
Calopogon pallidus, pale grass-pink
Listera smallii, Small's twayblade
Malaxis spicata, Florida adder's-mouth
Platanthera integrilabia, monkey-face orchis
Ponthieva racemosa, shadow-witch
Spiranthes sylvatica, woodland ladies'-tresses

or south into peninsular Florida:

Eulophia alta, wild coco
Habenaria odontopetala, toothed rein orchis
Habenaria quinqueseta, Michaux's orchid
Habenaria repens, water spider orchid
Mesadenus lucayanus, copper ladies'-tresses
Platanthera conspicua, southern white fringed orchis
Platythelys querceticola, low ground orchid, jug orchid
Pteroglossaspis ecristata, crestless plume orchid
Sacoila lanceolata, leafless beaked orchid
Spiranthes eatonii, Eaton's ladies'-tresses
Spiranthes floridana, Florida ladies'-tresses
Spiranthes longilabris, long-lipped ladies'-tresses
Spiranthes ovalis var. *ovalis,* southern oval ladies'-tresses
Triphora rickettii, Rickett's noddingcaps;

Spiranthes parksii, Navasota ladies'-tresses, is an exception as its primary range is west into east-central Texas

Four species and one variety of orchids have recently been described:

Calopogon oklahomensis, Oklahoma grass-pink 1995
Cypripedium kentuckiense, ivory-lipped lady's-slipper 1981
Spiranthes eatonii, Eaton's ladies'-tresses 1999
Spiranthes sylvatica, woodland ladies'-tresses 2001
Spiranthes ovalis var. *erostellata,* northern oval ladies'-tresses 1983

Eight new hybrids of orchids from the Southeast have recently been described:

Liparis ×jonesii 1999
Platanthera ×apalachicola 2003
Platanthera ×beckneri 2003
Platanthera ×lueri 2003
Platanthera ×osceola 2003
Spiranthes ×folsomii 2000
Spiranthes ×itchetuckneensis 1999
Spiranthes ×meridionalis 1999

Eighteen new forms of orchids have recently been described from the Southeast:

Calopogon barbatus forma *albiflorus*—white-flowered form 2003
forma *lilacinus*—lilac-flowered form 2003
Calopogon oklahomensis forma *albiflorus*—white-flowered form 2003
Cleistes divaricata forma *leucantha*—white-flowered form 1995

Corallorhiza wisteriana forma *rubra*—red-stemmed form 2000
Cypripedium kentuckiense forma *pricei*—white-flowered form 1998
forma *summersii*—concolorous yellow-flowered form 2003
Eulophia alta forma *pallida*—pale-flowered form 1995
forma *pelchatii*—white-flowered form 1998
Habenaria odontopetala forma *heatonii*—albino form 2001
Listera australis forma *scottii*—many-leaved form 2000
forma *viridis*—green-flowered form 2000
Malaxis spicata forma *trifoliata* 2003
Pteroglossaspis ecristata forma *flava*—yellow-flowered form 2000
forma *purpurea*—purple-flowered form 2003
Sacoila lanceolata var. *lanceolata* forma *folsomii*—golden bronze-flowered form 1999
Triphora trianthophora forma *caerulea*—blue-flowered form 2001
forma *rossii*—multi-colored form 1998

Four species and two varieties of orchids have recently been revalidated or taxonomically clarified:
Cleistes bifaria, upland spreading pogonia 1992
Corallorhiza odontorhiza var. *pringlei,* Pringle's autumn coralroot 1997
Cypripedium parviflorum var. *parviflorum,* southern small yellow lady's-slipper 1994
Malaxis bayardii, Bayard's adder's-mouth 1991
Platanthera conspicua, southern white fringed orchis 2002
Spiranthes floridana, Florida ladies'-tresses 1998

Three species of orchids are federally listed:
Isotria medeoloides, small whorled pogonia—threatened
Platanthera leucophaea, eastern prairie fringed orchis—threatened
Spiranthes parksii, Navasota ladies'-tresses—endangered

Ten additional species of orchids are among the rarest orchids in North America and considered globally threatened:
Calopogon multiflorus, many-flowered grass-pink
Cypripedium candidum, small white lady's-slipper
Gymnadeniopsis integra, yellow fringeless orchis
Malaxis bayardii, Bayard's adder's-mouth
Platanthera chapmanii, Chapman's fringed orchis
Platanthera integrilabia, monkey-face orchis

Spiranthes brevilabris, short-lipped ladies'-tresses, Texas ladies'-tresses
Spiranthes floridana, Florida ladies'-tresses
Spiranthes ovalis var. *ovalis*, southern oval ladies'-tresses
Triphora rickettii, Rickett's noddingcaps

Three species of orchids have recently been recognized in *Gymnadeniopsis* rather than in the genus *Habenaria* or *Platanthera*:
Gymnadeniopsis clavellata, little club-spur orchis
Gymnadeniopsis integra, yellow fringeless orchis
Gymnadeniopsis nivea, snowy orchis

One species of orchid has recently been recognized with a different species name than that which was traditionally used:
Epidendrum magnoliae, green-fly orchis = *E. conopseum*

Eight species of orchids range south of the United States, reaching the northern limit of their range in the southeastern United States:
Epidendrum magnoliae, green-fly orchis
Eulophia alta, wild coco
Habenaria odontopetala, toothed rein orchis
Habenaria quinqueseta, Michaux's orchid
Mesadenus lucayanus, copper ladies'-tresses
Platythelys querceticola, low ground orchid, jug orchid
Pteroglossaspis ecristata, crestless plume orchid
Sacoila lanceolata, leafless beaked orchid

Twenty-five species of orchids have their distribution primarily in the east-central and northeastern portions of North America and reach the southern limit of their range in the southeastern United States:
Arethusa bulbosa, dragon's-mouth
Calopogon tuberosus, common grass-pink
Corallorhiza maculata, spotted coralroot
Cypripedium acaule, pink lady's-slipper, moccasin flower
Epipactis helleborine, broad-leaved helleborine*
Galearis spectabilis, showy orchis
Goodyera pubescens, downy rattlesnake orchis
Gymnadeniopsis clavellata, little club-spur orchis
Isotria medeoloides, small whorled pogonia
Isotria verticillata, large whorled pogonia
Liparis loeselii, Löesel's twayblade, fen orchis
Listera australis, southern twayblade
Listera smallii, Small's twayblade

Malaxis bayardii, Bayard's adder's-mouth
Malaxis unifolia, green adder's-mouth
Platanthera blephariglottis, northern white fringed orchis
Platanthera flava var. *herbiola,* northern tubercled orchis
Platanthera grandiflora, large purple fringed orchis
Platanthera lacera, green fringed orchis, ragged orchis
Platanthera psycodes, small purple fringed orchis
Pogonia ophioglossoides, rose pogonia
Spiranthes cernua, nodding ladies'-tresses
Spiranthes lacera var. *gracilis,* southern slender ladies'-tresses
Spiranthes lucida, shining ladies'-tresses
Spiranthes ochroleuca, yellow ladies'-tresses

Rare, Threatened, and Endangered Species of Orchids

Each state has a somewhat different system for listing rare, threatened, and endangered plants. For that reason it is not possible to apply uniform criteria throughout the southeastern states. Lists are updated at various intervals and new species can be slow to appear. For the most recent information you should contact the agency listed for each state. Usually coding indicates: S1 = endangered, S2 = threatened, S3 = rare, SH = historical, and SX = extirpated. A different system is used in South Carolina: rc = rare, sc = special concern, ft = federally threatened. Florida uses none of the above. The entries are taken from lists published by each state. The southeastern United States region (as defined in this work) includes only a portion of Texas, Arkansas, and North Carolina, and plants that are listed outside of our range are not included. Species may be omitted from state lists for several reasons. The two most frequent are incomplete research and supporting data and undocumented historical records. Current names used in this work are given in parentheses.

Alabama

Aplectrum hyemale, putty-root S2
Calopogon barbatus, bearded grass-pink S1
Calopogon multiflorus, many-flowered grass-pink S1
Corallorhiza wisteriana, Wister's coralroot S2
Cypripedium candidum, small white lady's-slipper S1
Cypripedium kentuckiense, ivory-lipped lady's-slipper S1
Epidendrum conopseum, green-fly orchis S2 (*Epidendrum magnoliae*)
Isotria verticillata, large whorled pogonia S2
Liparis liliifolia, lily-leaved twayblade S1

Liparis loeselii, Löesel's twayblade S1?
Listera australis, southern twayblade S2
Platanthera blephariglottis var. *conspicua,* southern white fringed orchis S1S2 (*Platanthera conspicua*)
Platanthera flava var. *flava,* southern tubercled orchis S2S3
Platanthera integra, yellow fringeless orchis S1S2 (*Gymnadeniopsis integra*)
Platanthera integrilabia, monkey-face orchis S2
Platanthera lacera, green fringed orchis, ragged orchis S2
Platanthera nivea, snowy orchis S2? (*Gymnadeniopsis nivea*)
Platanthera peramoena, purple fringeless orchis S1
Ponthieva racemosa, shadow-witch S2
Pteroglossaspis ecristata, crestless plume orchid S1
Spiranthes longilabris, long-lipped ladies'-tresses S1
Spiranthes lucida, shining ladies'-tresses S1
Spiranthes magnicamporum, Great Plains ladies'-tresses S3

Source: Inventory List of Rare, Threatened and Endangered Plants, Animals and Natural Communities of Alabama, June 1999, The Nature Conservancy, Alabama Natural Heritage Program http://www.heritage.tnc.org/nhp/us/al/

Arkansas

Calopogon oklahomensis, Oklahoma grass-pink S2
Calopogon tuberosus, common grass-pink S2
Cypripedium kentuckiense, ivory-lipped lady's-slipper S3
Habenaria repens, water spider orchid S2
Hexalectris spicata, crested coralroot S2
Liparis loeselii, Löesel's twayblade S1
Platanthera cristata, orange crested orchis S1S2
Platanthera flava, southern tubercled orchis S1S2
Platanthera nivea, snowy orchis SH (*Gymnadeniopsis nivea*)
Platanthera peramoena, purple fringeless orchis S2
Pogonia ophioglossoides, rose pogonia S2
Spiranthes magnicamporum, Great Plains ladies'-tresses S1
Spiranthes odorata, fragrant ladies'-tresses S1
Spiranthes praecox, giant ladies'-tresses
(*Spiranthes sylvatica*) woodland ladies'-tresses S1S2

Source: Arkansas Natural Heritage Commission, 1500 Tower Building, 323 Center Street, Little Rock, AR 72201
Cindy Osborne, data manager

Phone: 501-324-9762, fax: 501-324-9618, e-mail: cindy@dah.state.ar.us
Sept. 27, 1999
http://www.heritage.state.ar.us:80/nhc/heritage.html

Florida

Listed Endangered

Calopogon multiflorus, many-flowered grass-pink
Corallorhiza odontorhiza, autumn coralroot
Goodyera pubescens, downy rattlesnake orchid
Hexalectris spicata, crested coralroot
Isotria verticillata, whorled pogonia
Malaxis unifolia, green adder's-mouth orchid
Platanthera clavellata, green rein orchid (*Gymnadeniopsis clavellata*)
Platanthera integra, orange rein orchid (*Gymnadeniopsis integra*)
Spiranthes brevilabris, small ladies'-tresses (including *S. floridana*)
Spiranthes ovalis, lesser ladies'-tresses
Spiranthes polyantha, Ft. George ladies'-tresses; does not grow in Florida; intent is to protect *Mesadenus lucayanus*

Listed Threatened

Cleistes divaricata, spreading pogonia: Proposed Endangered
Listera australis, southern twayblade
Platanthera blephariglottis, white-fringed orchid; intent is to protect *P. conspicua*
Platanthera ciliaris, yellow-fringed orchid
Platanthera cristata, crested fringed orchid
Platanthera flava, gypsy-spikes
Platanthera nivea, snowy orchid (*Gymnadeniopsis nivea*)
Pogonia ophioglossoides, rose pogonia
Pteroglossaspis ecristata, non-crested Eulophia: Proposed Endangered
Spiranthes laciniata, lace-lip ladies' tresses
Spiranthes longilabris, long-lip ladies' tresses: Proposed Endangered
Spiranthes tuberosa, little pearl-twist
Stenorrhynchos lanceolatum, leafless beaked orchid (*Sacoila lanceolata*)
Tipularia discolor, crane-fly orchid
Triphora trianthophora, three-birds orchid

Commercially Exploited

Epidendrum conopseum, green-fly orchid (*Epidendrum magnoliae*)

Proposed Endangered

Cleistes divaricata, large spreading pogonia
Platanthera chapmanii, Chapman's fringed orchis
Platythelys querceticola, low ground orchid
Pteroglossaspis ecristata, crestless plume orchid
Spiranthes floridana, Florida ladies'-tresses
Spiranthes longilabris, long-lipped ladies'-tresses
Triphora rickettii, Rickett's noddingcaps

Proposed Threatened

Cleistes bifaria, upland spreading pogonia
Spiranthes eatonii, Eaton's ladies'-tresses

Source: Florida Department of Agriculture and Consumer Services, Department of Plant Industry http://doacs.state.fl.us/~pi/5B-40.htm.

Georgia

Calopogon multiflorus, many-flowered grass-pink SH
Cleistes bifaria, upland spreading pogonia S1
Corallorhiza maculata, spotted coralroot SH
Cypripedium acaule, pink lady's-slipper, moccasin flower S4
Cypripedium calceolus var. *parviflorum,* southern small yellow lady's-slipper S2 (*Cypripedium parviflorum* var. *parviflorum*)
Cypripedium calceolus var. *pubescens,* large yellow lady's-slipper S3 (*Cypripedium parviflorum* var. *parviflorum*)
Epidendrum conopseum, green-fly orchis S3 (*Epidendrum magnoliae*)
Habenaria quinqueseta, Michaux's orchid S1
Isotria medeoloides, small whorled pogonia S2
Listera australis, southern twayblade S2
Listera smallii, Small's twayblade S2
Malaxis spicata, Florida adder's-mouth S1
Platanthera flava var. *herbiola,* northern tubercled orchis S1
Platanthera grandiflora, large purple fringed orchis S1
Platanthera integra, yellow fringeless orchis S2 (*Gymnadeniopsis integra*)
Platanthera integrilabia, monkey-face orchis S1S2
Platanthera nivea, snowy orchis S3 (*Gymnadeniopsis nivea*)
Platanthera peramoena, purple fringeless orchis S1
Platanthera psycodes, small purple fringed orchis S1?
Ponthieva racemosa, shadow-witch S2?
Pteroglossaspis ecristata, crestless plume orchid S1

Spiranthes brevilabris var. *floridana,* Florida ladies'- tresses S1 (*Spiranthes floridana*)

Spiranthes longilabris, long-lipped ladies'-tresses S1

Spiranthes magnicamporum, Great Plains ladies'-tresses S1

Spiranthes ovalis, southern oval ladies'-tresses S3

Source: Georgia Natural Heritage Program, 2117 U.S. Hwy. 278 S.E., Social Circle, GA 30025
Voice: 770-918-6411, 706-557-3032; fax: 706-557-3033
e-mail: greg_krakow@mail.dnr.state.ga.us
http://www.dnr.state.ga.us/dnr/wild/natural.html http://www.dnr.state.ga.us/dnr/wild/sppl_t.htm http://www.dnr.state.ga.us/dnr/wild/sppl_w.htm

Louisiana

Calopogon barbatus, bearded grass-pink s1

Calopogon multiflorus, many-flowered grass-pink s1

Calopogon pallidus, pale grass-pink 1s2

Cleistes divaricata, upland spreading pogonia s1 (*Cleistes bifaria*)

Corallorhiza odontorhiza, autumn coralroot s1

Cypripedium kentuckiense, ivory-lipped lady's-slipper s1

Habenaria quinqueseta, Michaux's orchid s1

Isotria verticillata, large whorled pogonia s2s3

Platanthera blephariglottis var. *conspicua,* southern white fringed orchis s1 (*Platanthera conspicua*)

Platanthera integra, yellow fringeless orchis s2s3 (*Gymnadeniopsis integra*)

Platanthera lacera, green fringed orchis, ragged orchis s1

Platythelys querceticola, low ground orchid, jug orchid s1

Pteroglossaspis ecristata, crestless plume orchid s2

Spiranthes magnicamporum, Great Plains ladies'-tresses s1

Triphora trianthophora, three birds orchid s1

Source: David Brunet
Brunet_DP@wlf.state.la.us

Mississippi

Aplectrum hyemale, putty-root s1

Calopogon barbatus, bearded grass-pink 2s3

Cleistes divaricata, upland spreading pogonia s3 (*Cleistes bifaria*)

Cypripedium pubescens, large yellow lady's-slipper s2s3 (*Cypripedium parviflorum* var. *pubescens*)

Cypripedium kentuckiense, ivory-lipped lady's-slipper su
Epidendrum conopseum, green-fly orchis s2 (*Epidendrum magnoliae*)
Erythrodes querceticola, low ground orchid, jug orchid s1? (*Platythelys querceticola*)
Eulophia ecristata, crestless plume orchid s1s2 (*Pteroglossaspis ecristata*)
Goodyera pubescens, downy rattlesnake orchis s1
Hexalectris spicata, crested coralroot s2
Orchis spectabilis, showy orchis s1 (*Galearis spectabilis*)
Platanthera blephariglottis, southern white fringed orchis s2 (including *Platanthera conspicua*)
Platanthera cristata, orange crested orchis s3
Platanthera integra, yellow fringeless orchis s3s4 (*Gymnadeniopsis integra*)
Platanthera integrilabia, monkey-face orchis s1
Platanthera lacera, green fringed orchis, ragged orchis s1s2
Platanthera peramoena, purple fringeless orchis s2s3
Ponthieva racemosa, shadow-witch s2?
Spiranthes longilabris, long-lipped ladies'-tresses s2s3
Spiranthes magnicamporum, Great Plains ladies'-tresses s2s3
Spiranthes ovalis, southern oval ladies'-tresses s2s3
Triphora trianthophora, three birds orchid s2s3

Source: Mississippi Natural Heritage Program
Mississippi Museum of Natural Science, 111 N. Jefferson, Jackson, MS 39202
Ronald Wieland, ecologist
heritage@mmns.state.ms.us

North Carolina

Calopogon multiflorus, many-flowered grass-pink s1
Cleistes bifaria, upland spreading pogonia s2?
Corallorrhiza wisteriana, Wister's coralroot s2
Cypripedium parviflorum, southern small yellow lady's-slipper s3 (*Cypripedium parviflorum* var. *parviflorum*)
Cypripedium pubescens, large yellow lady's-slipper s3 (*Cypripedium parviflorum* var. *pubescens*)
Habenaria repens, water spider orchid s2
Hexalectris spicata, crested coralroot s2
Isotria medeoloides, small whorled pogonia s1
Liparis loeselii, Löesel's twayblade s1
Listera australis, southern twayblade s3
Malaxis bayardii, Bayard's adder's-mouth sh

Malaxis spicata, Florida adder's-mouth s1
Platanthera flava var. *herbiola*, northern tubercled orchis s1?
Platanthera grandiflora, large purple fringed orchis s2
Platanthera integra, yellow fringeless orchis s1 (*Gymnadeniopsis integra*)
Platanthera integrilabia, monkey-face orchis sx
Platanthera nivea, snowy orchis s1 (*Gymnadeniopsis nivea*)
Platanthera peramoena, purple fringeless orchis s1
Ponthieva racemosa, shadow-witch s2
Pteroglossaspis ecristata, crestless plume orchid s1
Spiranthes brevilabris var. *floridana*, Florida ladies'-tresses sx? (*Spiranthes floridana*)
Spiranthes laciniata, lace-lipped ladies'-tresses s1
Spiranthes longilabris, long-lipped ladies'-tresses s1
Triphora trianthophora, three birds orchid s2?

Source: North Carolina Natural Heritage Program
http://ils.unc.edu/parkproject/nhp/2002plist.pdf

South Carolina

Arethusa bulbosa, dragon's-mouth rc
Calopogon barbatus, bearded grass-pink sc
Cypripedium pubescens, large yellow lady's-slipper sc (*Cypripedium parviflorum* var. *pubescens*)
Epidendrum conopseum, green-fly orchis sc (*Epidendrum magnoliae*)
Galearis spectabilis, showy orchis sc
Habenaria quinqueseta, Michaux's orchid sc
Isotria medeoloides, small whorled pogonia ft
Liparis liliifolia, lily-leaved twayblade sc
Listera australis, southern twayblade sc
Listera smallii, Small's twayblade sc
Platanthera integra, yellow fringeless orchis sc (*Gymnadeniopsis integra*)
Platanthera integrilabia, monkey-face orchis c2
Platanthera lacera, green fringed orchis, ragged orchis sc
Platanthera peramoena, purple fringeless orchis rc
Ponthieva racemosa, shadow-witch sc
Pteroglossaspis ecristata, crestless plume orchid c2
Spiranthes laciniata, lace-lipped ladies'-tresses sc
Spiranthes longilabris, long-lipped ladies'-tresses sc
Triphora trianthophora, three birds orchid sc

Source: Julie Holling
SCHPhttp://www.natureserve.org/nhp/us/sc/speclist.htm

Texas

Spiranthes parksii, Navasota ladies'-tresses FE

Source: Texas Parks and Wildlife Department,4200 Smith School Road, Austin, TX 78744
http://www.tpwd.state.tx.us/nature/endang/regulations/texas/

Recent Literature References for New Taxa, Combinations, and Additions to the Orchid Flora of the Southeastern United States

Two recent publications by Brown & Folsom, *Wild Orchids of Florida* (2002) and *The Wild Orchids of North America North of Mexico* (2003), have detailed the many recent publications that deal with new taxa, combinations, and additions to the orchid flora of the Southeast. In addition, the completion of the Orchidaceae for volume 26 of the *Flora of North America* (2002) also contains much additional information on all of the species in the Southeast. The reader is encouraged to consult these publications for more information. The following will enable the reader to easily reference these publications, as well as several additional new taxa that have been recently published.

Calopogon barbatus forma *albiflora* P. M. Brown
Brown, P. M. 2003. *North American Native Orchid Journal* 9: 33.
Calopogon barbatus forma *lilacinus* P. M. Brown
Brown, P. M. 2003. *North American Native Orchid Journal* 9: 34.
Calopogon multiflorus
Goldman, D. H. and S. Orzell. 2000. *Lindleyana* 15(4): 237–51.
Calopogon oklahomensis D. H. Goldman
Brown, P. M. 1995. *North American Native Orchid Journal* 1(2): 133.
Goldman, D. H. 1995. *Lindleyana* 10(1): 37–42.
Calopogon oklahomensis forma *albiflorus* P. M. Brown
Brown, P. M. 2003. *North American Native Orchid Journal* 9: 33–34.

Cleistes bifaria (Fernald) Catling & Gregg
Catling, P. M. and K. B. Gregg. 1992. *Lindleyana* 7(2): 57–73.
Cleistes divaricata forma *leucantha* P. M. Brown
Brown, P. M. 1995. *North American Native Orchid Journal* 1(1): 7.

Corallorhiza maculata (Rafinesque) Rafinesque var. *maculata* forma *flavida* (Peck) Farwell; forma *rubra* P. M. Brown
Brown, P. M. 1995. *North American Native Orchid Journal* 1(1): 8.
Corallorhiza odontorhiza (Willdenow) Nuttall var. *pringlei* (Greenman) Freudenstein
Freudenstein, J. V. 1993. Dissertation. Cornell University.
———. 1997. *Harvard Papers in Botany* 10: 5–51.
Corallorhiza wisteriana forma *albolabia* P. M. Brown
Brown, P. M. 1995. *North American Native Orchid Journal* 1(1): 9–10.
Corallorhiza wisteriana forma *rubra* P. M. Brown
Brown, P. M. 2000. *North American Native Orchid Journal* 6(1): 62.

Cypripedium acaule forma *biflorum* P. M. Brown
Brown, P. M. 1995. *North American Native Orchid Journal* 1(3): 197.
Cypripedium kentuckiense C. F. Reed
Atwood, J. T., Jr. 1984. *AOS Bulletin* 53(8): 835–41.
Brown, P. M. 1995. *North American Native Orchid Journal* 1(3): 255.
Reed, C. 1981. *Phytologia* 48(5): 426–28.
Weldy, T. W., H. T. Mlodozeniec, L. E. Wallace, and M. A. Case. 1996. *Sida* 17(2): 423–35.
Cypripedium kentuckiense forma *pricei* P. M. Brown
Brown, P. M. 1998. *North American Native Orchid Journal* 4(1): 45.
Cypripedium kentuckiense forma *summersii* P. M. Brown
Brown, P. M. 2002. *North American Native Orchid Journal* 8: 30–31.
Cypripedium parviflorum Salisbury var. *parviflorum*
Sheviak, C. J. 1994. *AOS Bulletin* 63(6): 664–69.
———. 1995. *AOS Bulletin* 64(6): 606–12.
———. 1996. *North American Native Orchid Journal* 2(4): 319–43.
Cypripedium parviflorum Salisbury var. *parviflorum* forma *albolabium* Magrath & Norman
Magrath, L. K. and J. L. Norman. 1989. *Sida* 13(3): 371–72.
Cypripedium parviflorum Salisbury var. *pubescens* (Willdenow) Knight
Sheviak, C. J. 1994. *AOS Bulletin* 63(6): 664–69.
———. 1995. *AOS Bulletin* 64(6): 606–12.
———. 1996. *North American Native Orchid Journal* 2(4): 319–43.

Epidendrum magnoliae Mühlenberg
Hágsater, E. 2000. *North American Native Orchid Journal* 6(4): 299–309.

Epipactis helleborine forma *luteola* P. M. Brown
Brown, P. M. 1996. *North American Native Orchid Journal* 2(4): 316.

Eulophia alta forma *pallida* P. M. Brown
Brown, P. M. 1995. *North American Native Orchid Journal* 1(2): 131.
Eulophia alta forma *pelchatii* P. M. Brown
Brown, P. M. 1998. *North American Native Orchid Journal* 4(1): 46.

Galearis spectabilis forma *willeyi* (Seymour) P. M. Brown
Brown, P. M. 1988. *Wild Flower Notes* 3(1): 20.

Gymnadeniopsis clavellata (Michaux) Rydberg
Gymnadeniopsis integra (Nuttall) Rydberg
Gymnadeniopsis nivea (Nuttall) Rydberg
Brown, P. M. *North American Native Orchid Journal* 8:32–40.

Liparis ×jonesii S. Bentley
Bentley, S. 2000. *Native Orchids of the Southern Appalachian Mountains,* pp. 138–39.

Listera smallii forma *variegata* P. M. Brown
Brown, P. M. 1995. *North American Native Orchid Journal* 1(4): 289.

Malaxis bayardii Fernald
Catling, P. M. 1991. *Lindleyana* 6(1): 3–23.
Malaxis spicata forma *trifoliata* P. M. Brown
Brown, P. M. 2003. *North American Native Orchid Journal* 9: 34.

Mesadenus lucayanus (Britton) Schlechter
Brown, P. M. 2000. *North American Native Orchid Journal* 6(4): 333–34.

Platanthera chapmanii (Small) Luer *emend.* Folsom
Brown, P. M. and S. Stewart. 2003. *North American Native Orchid Journal* 9: 35–37.
Folsom, J. P. 1984. *Orquidea* (Mex.) 9(2): 344.
Platanthera conspicua (Nash) P. M. Brown
Brown, P. M. 2002. *North American Native Orchid Journal* 8: 3–14.
Platanthera grandiflora (Bigelow) Lindley
Stoutamire, W. P. 1974. *Brittonia* 26: 42–58.
Platanthera grandiflora forma *bicolor* P. M. Brown; forma *carnea* P. M. Brown
Brown, P. M. 1995. *North American Native Orchid Journal* 1(1): 12.
Platanthera grandiflora forma *mentotonsa* (Fernald) P. M. Brown
Brown, P. M. 1988. *Wild Flower Notes* 3(1): 22.
Platanthera integrilabia (Correll) Luer
Ranger. L. S. 1994. *Tipularia* 9: 7–13.
Zettler, L. W. 1994. *AOS Bulletin* 63(6): 686–88.

Zettler, L. W. and J. E. Fairey III. 1992. *Lindleyana* 5(4): 212–17.
Platanthera leucophaea (Nuttall) Lindley
Sheviak, C. J. and M. Bowles. 1986. *Rhodora* 88: 267–90.
Platanthera peramoena (A. Gray) A. Gray
Spooner, D. M. and J. S. Shelly. 1983. *Rhodora* 85: 55–64.
Platanthera peramoena forma *doddiae* P. M. Brown
Brown, P. M. 2002. *North American Native Orchid Journal* 8: 30–31.
Platanthera psycodes (Linnaeus) Lindley
Stoutamire, W. P. 1974. *Brittonia* 26: 42–58.
Platanthera psycodes forma *ecalcarata* (Bryan) P. M. Brown; forma *varians* (Bryan) P. M. Brown
Brown, P. M. 1988. *Wild Flower Notes* 3(1): 24.
Platanthera psycodes forma *rosea* P. M. Brown
Brown, P. M. 1995. *North American Native Orchid Journal* 1(4): 289.
Platanthera ×*apalachicola* P. M. Brown & S. Stewart
Brown, P. M. and S. Stewart. 2003. *North American Native Orchid Journal* 9: 35.
Platanthera ×*beckneri* P. M. Brown
Brown, P. M. 2003. *North American Native Orchid Journal* 8: 3–14.
Platanthera ×*channellii* Folsom
Folsom, J. P. 1984. *Orquidea* (Mex.) 9(2): 344.
Platanthera ×*lueri* P. M. Brown
Brown, P. M. 2003. *North American Native Orchid Journal* 8: 3–14.
Platanthera ×*osceola* P. M. Brown & S. Stewart
Brown, P. M. and S. Stewart. 2003. *North American Native Orchid Journal* 9: 35.

Pogonia ophioglossoides forma *brachypogon* (Fernald) P. M. Brown
Brown, P. M. 1998. *North American Native Orchid Journal* 6(4): 339.

Pteroglossaspis ecristata forma *flava* P. M. Brown
Brown, P. M. 2000. *North American Native Orchid Journal* 6(1): 64.
Pteroglossaspis ecristata forma *purpurea* P. M. Brown
Brown, P. M. 2003. *North American Native Orchid Journal* 9: 34.

Sacoila lanceolata var. *lanceolata* forma *albidaviridis* Catling & Sheviak
Catling, P. M. and C. J. Sheviak. 1993. *Lindleyana.* 8(2): 77–81.
Sacoila lanceolata var. *lanceolata* forma *folsomii* P. M. Brown
Brown, P. M. 1999. *North American Native Orchid Journal* 5(2): 198.

Spiranthes cernua (Linnaeus) L. C. Richard
Sheviak, C. J. 1991. *Lindleyana* 6(4): 228–34.
Spiranthes eatonii Ames *ex* P. M. Brown

Brown, P. M. 1999. *North American Native Orchid Journal* 5(1): 5.
Spiranthes floridana (Cory) Wherry *emend.* P. M. Brown
Brown, P. M. 2001. *North American Native Orchid Journal* 7(1): 91–93.
Sorrie, B. A. 1998. *Sida* 18(3): 904.
Spiranthes lucida (H. H. Eaton) Ames
Allison, J. R. and T. E. Stevens. 2001. *Castanea* 66(1–2): 154–205.
Spiranthes ovalis Lindley var. *ovalis*
Spiranthes ovalis Lindley var. *erostellata* Catling
Catling, P. M. 1983. *Brittonia* 35: 120–25.
Spiranthes praecox forma *albolabia* Brown & McCartney
Brown, P. M. 1995. *North American Native Orchid Journal* 1(1): 13.
Spiranthes sylvatica P. M. Brown
Brown, P. M. 2001. *North American Native Orchid Journal* 7(3): 193–205.
Spiranthes ×*folsomii* P. M. Brown
Brown, P. M. 2000. *North American Native Orchid Journal* 6(1): 16.
Spiranthes ×*intermedia* Ames
Catling, P. M. 1978. *Rhodora* 80: 377–89.
Spiranthes ×*itchetuckneensis* P. M. Brown
Brown, P. M. 1999. *North American Native Orchid Journal* 5(4): 358–67.
Spiranthes ×*meridionalis* P. M. Brown
Brown, P. M. 1999. *North American Native Orchid Journal* 5(4): 358–67.
———. 2000. *North American Native Orchid Journal* 6(2): 139.

Tipularia discolor (Pursh) Nuttall forma *viridifolia* P. M. Brown
Brown, P. M. 2000. *North American Native Orchid Journal* 6(4): 334–35.

Triphora trianthophora forma *albidoflava* Keenan
Keenan, P. 1992. *Rhodora* 94: 38–39.
Triphora trianthophora forma *caerulea* P. M. Brown
Brown, P. M. 1999. *North American Native Orchid Journal* 7(1): 3.
Triphora trianthophora forma *rossii* P. M. Brown
Brown, P. M. 1999. *North American Native Orchid Journal* 5(1): 5.

Synonyms and Misapplied Names

Synonyms and misapplied names are often confused both in the literature and in the understanding of orchid enthusiasts. A synonym is simply an alternate name for a previously published plant name. From among the synonyms each author must select a name that they feel best suits the currently accepted genus and species for a given plant. Although the genus may vary, the species epithet may often be the same. In such large groups as the spiranthoid orchids (*Spiranthes* and its allied genera), many synonyms may exist for the same species—all within different genera.

The rules of priority, as set forth in the *International Code of Botanical Nomenclature* (2000), dictate that the earliest validly published name must be used. A good example would be that of the green-fly orchis, *Epidendrum magnoliae* Mühlenberg, published in October 1813, and *Epidendrum conopseum* R. Brown, published in November 1813. Although the latter name is in widespread usage, priority indicates that the former is actually the valid name. It rarely comes down to months, as in this example, and usually the year of publication is sufficient for determining the valid name.

Spiranthes lanceolata would be a synonym for *Sacoila lanceolata*—*Sacoila* being the currently accepted genus. *Habenaria* is another group that has undergone a great deal of scrutiny in the past twenty-five years. Several groups of species formerly included within *Habenaria* are now treated as distinct genera. This is not always so much a case of correct or incorrect names, but that of the preference of the author for one genus over another. A synonym may also be a validly published name that duplicated a previously published taxon and therefore is rendered a synonym.

A misapplied name is an incorrect name for a given plant that may have resulted from a reassessment of the genus or species, resulting in two or more species being described from within the original species, or it may simply be a wrong name assigned to the plants. This is especially common in geographic areas at the edge of a group's range. *Habenaria floribunda* would be a misapplied name for *H. odontopetala, H. floribunda* not occurring in Florida, or it may be an error in the original identification. A more frequently encountered example would be *Cypripedium calceolus.* For many years our North American yellow lady's-slippers have been treated as a geographic variant of the Eurasian species. Many years of work by Sheviak have demonstrated that this is not the case and that the North American plants are a distinct species—*Cypripedium parviflorum.* Therefore, the name *C. calceolus* is a misapplied name for the North American plants. The term *auct.* (*auctorum,* i.e., of authors) is used to indicate a misapplied name and occasionally an author will incorrectly append the phrase "in part" after a name listed under synonymy. Misapplied names are not synonyms and refer only to the specific geographic area being treated—in this case the southeastern United States.

An issue can arise as to whether a name is a synonym or misapplied, and that depends on a broad or narrow view of the taxonomy—the lumpers vs. the splitters. Such a situation would best be described thusly: if *Malaxis bayardii* is considered to be synonymous with *Malaxis unifolia, M. bayardii* becomes a synonym of *M. unifolia;* but if *M. bayardii* is considered a good species on its own, *M. unifolia* becomes a misapplied name for *M. bayardii.* At times this appears to be an endless argument and each author must make his or her own decision as to synonymy and misapplied names. In the southeastern United States there are very few misapplied names, whereas to the south, in peninsular Florida, there are many taxa to be found with misapplied names. Cross-references to all taxa with synonyms and misapplied names can be found at the end of this chapter.

Synonyms and misapplied names are given for taxa found in the following sources as well as in occasional references to specific journal articles:

Bentley, S. 2000. *Native Orchids of the Southern Appalachians Mountains.*

Brown, P. M. and S. N. Folsom. 2002. *Wild Orchids of Florida.*

———. 2003. *The Wild Orchids of North America, North of Mexico.*

Correll, D. S. 1950. *Native Orchids of North America.*

Liggio, J. and A. Liggio. 1999. *Wild Orchids of Texas.*

Luer, C. A. 1972. *The Native Orchids of Florida.*

———. 1975. *The Native Orchids of the United States and Canada excluding Florida.*

Flora of North America, vol. 26. *Orchidaceae.* 2002.

Slaughter, C. R. 1995. *Wild Orchids of Arkansas.*

Small, J. K. 1933. *Manual of Southeastern Flora.*

Wunderlin, R. P. 1998. *A Guide to the Vascular Plants of Florida.*

Segregate Genera

Three genera within the range of this book have several segregate genera that have been variously treated by different authors. Those are:

Habenaria
- *Coeloglossum*
- *Gymnadeniopsis*
- *Platanthera*

Pogonia
- *Cleistes*
- *Isotria*

Spiranthes
- *Mesadenus*
- *Sacoila*

Synonyms

Most current taxonomic treatments recognize the numerous segregate genera of the spiranthoid orchids (Garay, 1980). References given are for recent taxonomic treatments, not necessarily the taxonomic work that first designated the new combination. Volume 26 of the *Flora of North America* (2002) treats many of the following taxa as well.

Cleistes bifaria (Fernald) Catling & Gregg
SYNONYMS
Cleistes divaricata var. *bifaria* Fernald
Catling, P. M. and K. B. Gregg 1992. *Lindleyana* 7(2): 57–73.
Pogonia bifaria P. M. Brown & R. P. Wunderlin
Brown, P. M. and R. P. Wunderlin. 1997. *North American Native Orchid Journal* (3): 450–51.

Coeloglossum viride (Linnaeus) Hartman var. *virescens* (Mühlenberg) Luer
SYNONYMS
Coeloglossum bracteatum (Mühlenberg) *ex* Willdenow) Parlin
Coeloglossum viride ssp. *bracteatum* (Mühlenberg *ex* Willdenow) Hultén
Dactylorhiza viridis R. M. Bateman, A. Pridgeon & M. W. Chase
Habenaria bracteata (Mühlenberg *ex* Willdenow) R. Brown *ex* Aiton *f.*
Habenaria viridis Linnaeus var. *bracteata* (Mühlenberg *ex* Willdenow) Reichenbach *ex* Gray

Corallorhiza maculata (Rafinesque) Rafinesque var. *maculata*
SYNONYM
Corallorhiza multiflora Nuttall

Corallorhiza odontorhiza (Willdenow) Nuttall var. *pringlei* (Greenman) Freudenstein
SYNONYM
Corallorhiza pringlei Greenman

Cypripedium acaule Aiton
SYNONYM
Fissipes acaulis (Aiton) Small

Cypripedium kentuckiense C. F. Reed
SYNONYM
C. daultonii V. Soukup *nom. nud.*

Cypripedium parviflorum Salisbury var. *parviflorum*
SYNONYMS
Cypripedium calceolus Linnaeus var. *parviflorum* (Salisbury) Fernald *p.p.*
Cypripedium parviflorum Salisbury var. *pubescens* (Willdenow) Knight
Cypripedium pubescens Willdenow
Cypripedium calceolus L. var. *planipetalum* (Fernald) Victorin & Rousseau
Cypripedium calceolus Linnaeus var. *pubescens* (Willdenow) Correll
Cypripedium parviflorum var. *planipetalum* Fernald
Cypripedium flavescens de Candolle
Cypripedium veganum Cockerell & Barber
MISAPPLIED
Cypripedium calceolus Linnaeus

Cypripedium reginae Walter
SYNONYM
Cypripedium spectabile Salisbury

Epidendrum magnoliae Mühlenberg
SYNONYM
Epidendrum conopseum R. Brown
Hágsater, E. 2000. *North American Native Orchid Journal* 6(4): 299–309.

Epipactis helleborine (Linnaeus) Cranz*
SYNONYM
Epipactis latifolia (Linnaeus) Allioni*

Gymnadeniopsis clavellata (Michaux) Rydberg *ex* Britton
SYNONYMS
Platanthera clavellata (Michaux) Luer

Habenaria clavellata (Michaux) Sprengel
Brown, P. M. 2002. *North American Native Orchid Journal* 8: 32–40.

Gymnadeniopsis integra (Nuttall) Rydberg
SYNONYMS
Platanthera integra (Nuttall) Gray *ex* Beck
Habenaria integra (Nuttall) Sprengel
Brown, P. M. 2002. *North American Native Orchid Journal* 8: 32–40.

Gymnadeniopsis nivea (Nuttall) Rydberg
SYNONYMS
Platanthera nivea (Nuttall) Luer
Habenaria nivea (Nuttall) Sprengel
Brown, P. M. 2002. *North American Native Orchid Journal* 8: 32–40.

Habenaria odontopetala Reichenbach *f.*
SYNONYMS
Habenaria strictissima Reichenbach *f.* var. *odontopetala* (Reichenbach *f.*) L. O. Williams
Habenaria garberi Porter
MISAPPLIED
Habenaria floribunda Lindley

Wunderlin (1997) used *Habenaria floribunda,* considering it a polymorphic umbrella species for several other taxa. Sheviak (pers. comm.; *FNA* 2002) could find no evidence for including *H. odontopetala* within *H. floribunda.*

Isotria medeoloides (Pursh) Rafinesque
SYNONYMS
Pogonia affinis Austin *ex* A. Gray
Isotria affinis (Austin *ex* A. Gray) Rydberg

Isotria verticillata (Mühlenberg *ex* Willdenow) Rafinesque
SYNONYM
Pogonia verticillata (Mühlenberg *ex* Willdenow) Nuttall

Listera australis Lindley
SYNONYM
Neottia australis (Lindley) Szlachetko

Listera smallii Wiegand
SYNONYMS
Listera reniformis Small
Neottia smallii (Wiegand) Szlachetko

Mesadenus lucayanus (Britton) Schlechter
SYNONYMS
Ibidium lucayanum Britton
Spiranthes lucayana (Britton) Cogniaux
MISAPPLIED
Mesadenus polyanthus (Reichenbach *f.*) Schlechter

Recent examinations of numerous specimens of *Mesadenus polyanthus* from Mexico compared to the material from the West Indies and Florida clearly indicate that two species are present. *Mesadenus polyanthus* is confined to higher elevations in central Mexico and *M. lucayanus* to lower elevations in Florida, the West Indies, southern Mexico, and Guatemala.

Brown, P. M. 2000. *North American Native Orchid Journal* 6(4): 333–34.

Platanthera blephariglottis (Willdenow) Lindley var. *blephariglottis*
SYNONYMS
Blephariglottis blephariglottis (Willdenow) Rydberg
Habenaria blephariglottis (Willdenow) Hooker

Platanthera chapmanii (Small) Luer *emend.* Folsom
SYNONYMS
Blephariglottis chapmanii Small
Habenaria ×*chapmanii* (Small) Ames
Folsom, J. P. 1984. *Orquidea* (Mex.) 9(2): 344.

Platanthera ciliaris (Linnaeus) Lindley
SYNONYMS
Blephariglottis ciliaris (Linnaeus) Rydberg
Habenaria ciliaris (Linnaeus) R. Brown

Platanthera conspicua (Nash) P. M. Brown
SYNONYMS
Platanthera blephariglottis (Willdenow) Lindley var. *conspicua* (Nash) Luer
Blephariglottis conspicua (Nash) Small
Habenaria blephariglottis (Willdenow) Hooker var. *conspicua* (Nash) Ames
Brown, P. M. 2002. *North American Native Orchid Journal* 8: 3–14.

Platanthera cristata (Michaux) Lindley
SYNONYMS
Blephariglottis cristata (Michaux) Rafinesque
Habenaria cristata (Michaux) R. Brown

Platanthera flava (Linnaeus) Lindley var. *flava*
SYNONYM
Habenaria flava (Linnaeus) R. Brown

Platanthera flava (Linnaeus) Lindley var. *herbiola* (R. Brown) Luer
SYNONYMS
Habenaria herbiola R. Brown *ex* Aiton
Habenaria flava var. *herbiola* (R. Brown *ex* Aiton) Ames & Correll
Habenaria flava var. *virescens sensu* Fernald

Platanthera grandiflora (Bigelow) Lindley
SYNONYMS
Blephariglottis grandiflora (Bigelow) Rydberg
Habenaria fimbriata (Dryander) R. Brown *ex* Aiton
Habenaria grandiflora (Bigelow) Torrey
Habenaria psycodes (Linnaeus) Sprengel var. *grandiflora* (Bigelow) A. Gray

Platanthera integrilabia (Correll) Luer
SYNONYMS
Habenaria blephariglottis (Willdenow) Hooker var. *integrilabia* Correll
Habenaria correlliana Cronquist

Platanthera lacera (Michaux) G. Don
SYNONYM
Habenaria lacera (Michaux) R. Brown

Platanthera leucophaea (Nuttall) Lindley
SYNONYM
Habenaria leucophaea (Nuttall) A. Gray

Platanthera peramoena (A. Gray) A. Gray
SYNONYM
Blephariglottis peramoena (Gray) Rydberg
Habenaria peramoena A. Gray

Platanthera psycodes (Linnaeus) Lindley
SYNONYM
Blephariglottis psycodes (Linnaeus) Rydberg
Habenaria psycodes (Linnaeus) Sprengel

Platanthera ×bicolor (Rafinesque) Luer
SYNONYM
Habenaria ×bicolor (Rafinesque) Beckner

Platanthera ×*canbyi* (Ames) Luer
SYNONYMS
Blephariglottis canbyi (Ames) W. Stone
Habenaria canbyi Ames

Platythelys querceticola (Lindley) Garay
SYNONYMS
Erythrodes querceticola (Lindley) Ames
Physurus querceticola Lindley

Malaxis spicata Swartz
SYNONYM
Malaxis floridana (Chapman) O. Kuntze

Pteroglossaspis ecristata (Grisebach) Rolfe
SYNONYM
Eulophia ecristata (Fernald) Ames

Sacoila lanceolata (Aublet) Garay var. *lanceolata*
SYNONYMS
Spiranthes lanceolata (Aublet) Leon
Spiranthes orchioides (Swartz) A. Richard
Stenorrhynchos lanceolatum (Aublet) Richard *ex* Sprengel

All of the *Spiranthes* synonyms have had a long and somewhat checkered history. Small used the genus *Ibidium* for virtually all of the *Spiranthes* he treated and Correll tended to merge many of the segregate genera as well as several species in his 1950 publication. It has been a long and slow process to re-evaluate and correctly address this situation. Although Garay (1980) both restored and designated some new genera, many of the following have not appeared in current literature, other than a few journal articles, until the publication of *The Wild Orchids of North America, North of Mexico* (2003) and volume 26 of *Flora of North America* (2002).

Spiranthes brevilabris Lindley
SYNONYM
Spiranthes gracilis (Bigelow) L. C. Beck var. *brevilabris* (Lindley) Correll

Spiranthes cernua (Linnaeus) Richard
SYNONYM
Ibidium cernuum (Linnaeus) House

Spiranthes floridana (Wherry) Cory *emend.* P. M. Brown
SYNONYMS
Ibidium floridanum Wherry
Spiranthes brevilabris Lindley var. *floridanum* (Wherry) Luer
Spiranthes gracilis (Bigelow) L. C. Beck var. *floridanum* (Wherry) Correll

Spiranthes lacera Rafinesque var. *gracilis* (Bigelow) Luer
SYNONYMS
Neottia gracilis Bigelow
Spiranthes gracilis (Bigelow) Beck
Spiranthes beckii Lindley
Ibidium gracile (Bigelow) House
Ibidium beckii (Lindley) House

Spiranthes laciniata (Small) Ames
SYNONYM
Ibidium laciniatum (Small) House

Spiranthes longilabris Lindley
SYNONYM
Ibidium longilabre (Lindley) House

Spiranthes ochroleuca (Rydberg) Rydberg
SYNONYMS
Gyrostachys ochroleuca Rydberg *ex* Britton
Spiranthes cernua (Linnaeus) L. C. Richard var. *ochroleuca* (Rydberg) Ames
Spiranthes ×*steigeri* Correll

Spiranthes odorata (Nuttall) Lindley
SYNONYMS
Spiranthes cernua (Linnaeus) L. C. Richard var. *odorata* (Nuttall) Correll
Ibidium odoratum (Nuttall) House

Spiranthes praecox (Walter) S. Watson
SYNONYM
Ibidium praecox (Walter) House

Spiranthes tuberosa Rafinesque
SYNONYMS
Spiranthes grayi Ames
Spiranthes tuberosa var. *grayi* (Ames) Fernald
Spiranthes beckii Lindley

MISAPPLIED
Ibidium beckii House

Spiranthes vernalis Englemann & A. Gray
SYNONYM
Ibidium vernale (Englemann & A. Gray) House

Triphora rickettii Luer
MISAPPLIED
Triphora yucatanensis Ames
Although Medley (*Selbyana* 12: 102–3, 1991) considered these two species the same, detailed examination of *Triphora rickettii* in the field has shown several differences, and I feel that it should be maintained as a separate species.

Cross References for Synonyms and Misapplied Names

The synonym is given first, followed by the currently acceptable name.
= synonym
≠ misapplied name

Blephariglottis blephariglottis (Willdenow) Rydberg = *Platanthera blephariglottis* (Willdenow) Lindley var. *blephariglottis*
Blephariglottis canbyi (Ames) W. Stone = *Platanthera* ×*canbyi* (Ames) Luer
Blephariglottis chapmanii Small = *Platanthera chapmanii* (Small) Luer *emend.* Folsom
Blephariglottis ciliaris (Linnaeus) Rydberg = *Platanthera ciliaris* (Linnaeus) Lindley
Blephariglottis conspicua (Nash) Small = *Platanthera conspicua* (Nash) P. M. Brown
Blephariglottis cristata (Michaux) Rafinesque = *Platanthera cristata* (Michaux) Lindley
Blephariglottis grandiflora (Bigelow) Rydberg = *Platanthera grandiflora* (Bigelow) Lindley
Blephariglottis peramoena (Gray) Rydberg = *Platanthera peramoena* (A. Gray) A. Gray
Blephariglottis psycodes (Linnaeus) Rydberg = *Platanthera psycodes* (Linnaeus) Lindley
Calopogon pulchellus (Salisbury) R. Brown = *Calopogon tuberosus* (Linnaeus) Britton, Sterns, & Poggenberg
Cleistes divaricata var. *bifaria* Fernald = *Cleistes bifaria* (Fernald) Catling & Gregg
Coeloglossum bracteatum (Mühlenberg) *ex* Willdenow) Parlin = *Coeloglossum viride* (Linnaeus) Hartman var. *virescens* (Mühlenberg) Luer

Coeloglossum viride ssp. *bracteatum* (Mühlenberg *ex* Willdenow) Hultén = *Coeloglossum viride* (Linnaeus) Hartman var. *virescens* (Mühlenberg) Luer

Corallorhiza multiflora Nuttall = *Corallorhiza maculata* (Rafinesque) Rafinesque

Corallorhiza pringlei Greenman = *Corallorhiza odontorhiza* (Willdenow) Poiret var. *pringlei* (Greenman) Freudenstein

Cypripedium calceolus Linnaeus var. *planipetalum* (Fernald) Victorin & Rousseau = ***Cypripedium parviflorum*** Salisbury var. *pubescens* (Willdenow) Knight

Cypripedium calceolus Linnaeus ≠ *Cypripedium parviflorum* Salisbury var. *pubescens* (Willdenow) Knight

Cypripedium calceolus Linnaeus var. *parviflorum* (Salisbury) Fernald *p.p.* = *Cypripedium parviflorum* Salisbury var. *parviflorum*

Cypripedium calceolus Linnaeus var. *pubescens* (Willdenow) Correll = *Cypripedium parviflorum* Salisbury var. *pubescens* (Willdenow) Knight

Cypripedium daultonii V. Soukup *nom. nud.* = *C. kentuckiense* C. F. Reed

Cypripedium flavescens de Candolle = *Cypripedium parviflorum* Salisbury var. *pubescens* (Willdenow) Knight

Cypripedium parviflorum var. *planipetalum* Fernald = *Cypripedium parviflorum* Salisbury var. *pubescens* (Willdenow) Knight

Cypripedium pubescens Willdenow = *Cypripedium parviflorum* Salisbury var. *pubescens* (Willdenow) Knight

Cypripedium spectabile Salisbury = *Cypripedium reginae* Walter

Cypripedium veganum Cockerell & Barber = *Cypripedium parviflorum* Salisbury var. *pubescens* (Willdenow) Knight

Dactylorhiza viridis (Linnaeus) R.M. Bateman, A. Pridgeon, & M.W. Chase = *Coeloglossum viride* Linnaeus

Epidendrum conopseum R. Brown = *Epidendrum magnoliae* Mühlenberg

Epipactis latifolia (Linnaeus) Allioni* = *Epipactis helleborine* (Linnaeus) Cranz*

Erythrodes querceticola (Lindley) Ames = *Platythelys querceticola* (Lindley) Garay

Eulophia ecristata (Fernald) Ames = *Pteroglossaspis ecristata* (Grisebach) Rolfe

Fissipes acaulis (Aiton) Small = *Cypripedium acaule* Aiton

Galeorchis spectabilis (Linnaeus) Rydberg = *Galearis spectabilis* (Linnaeus) Rafinesque

Gyrostachys ochroleuca Rydberg *ex* Britton = *Spiranthes ochroleuca* (Rydberg) Rydberg

Gyrostachys laciniata Small = *Spiranthes laciniata* (Small) Ames

Habenaria ×*bicolor* (Rafinesque) Beckner = *Platanthera* ×*bicolor* (Rafinesque) Luer

Habenaria ×*chapmanii* (Small) Ames = *Platanthera chapmanii* (Small) Luer *emend.* Folsom

Habenaria blephariglottis (Willdenow) Hooker = *Platanthera blephariglottis* (Willdenow) Lindley var. *blephariglottis*

Habenaria blephariglottis (Willdenow) Hooker var. *conspicua* (Nash) Ames = *Platanthera conspicua* (Nash) P. M. Brown
Habenaria blephariglottis (Willdenow) Hooker var. *integrilabia* Correll = *Platanthera integrilabia* (Correll) Luer
Habenaria bracteata (Mühlenberg *ex* Willdenow) R. Brown *ex* Aiton *f.* = *Coeloglossum viride* (Linnaeus) Hartman var. *virescens* (Mühlenberg) Luer
Habenaria canbyi Ames = *Platanthera* ×*canbyi* (Ames) Luer
Habenaria ciliaris (Linnaeus) R. Brown = *Platanthera ciliaris* (Linnaeus) Lindley
Habenaria clavellata (Michaux) Sprengel = *Gymnadeniopsis clavellata* (Michaux) Rydberg *ex* Britton
Habenaria correlliana Cronquist = *Platanthera integrilabia* (Correll) Luer
Habenaria cristata (Michaux) R. Brown = *Platanthera cristata* (Michaux) Lindley
Habenaria fimbriata (Dryander) R. Brown *ex* Aiton = *Platanthera grandiflora* (Bigelow) Lindley
Habenaria flava (Linnaeus) R. Brown = *Platanthera flava* (Linnaeus) Lindley var. *flava*
Habenaria flava (Linnaeus) R. Brown var. *herbiola* (R. Brown *ex* Aiton) Ames & Correll = *Platanthera flava* (Linnaeus) Lindley var. *herbiola* (R. Brown) Luer
Habenaria flava (Linnaeus) R. Brown var. *virescens sensu* Fernald = *Platanthera flava* (Linnaeus) Lindley var. *herbiola* (R. Brown) Luer
Habenaria floribunda Lindley ≠ *Habenaria odontopetala* Reichenbach *f.*
Habenaria garberi Porter = *Habenaria odontopetala* Reichenbach *f.*
Habenaria grandiflora (Bigelow) Torrey = *Platanthera grandiflora* (Bigelow) Lindley
Habenaria herbiola R. Brown *ex* Aiton = *Platanthera flava* (Linnaeus) Lindley var. *herbiola* (R. Brown) Luer
Habenaria integra (Nuttall) Sprengel = *Gymnadeniopsis integra* Nuttall
Habenaria lacera (Michaux) R. Brown = *Platanthera lacera* (Michaux) G. Don
Habenaria leucophaea (Nuttall) A. Gray = *Platanthera leucophaea* (Nuttall) Lindley
Habenaria nivea (Nuttall) Sprengel = *Gymnadeniopsis nivea* Nuttall
Habenaria nuttallii Small = *Habenaria repens* Nuttall
Habenaria peramoena A. Gray = *Platanthera peramoena* (A. Gray) A. Gray
Habenaria psycodes (Linnaeus) Sprengel = *Platanthera psycodes* (Linnaeus) Lindley
Habenaria psycodes (Linnaeus) Sprengel var. *grandiflora* (Bigelow) A. Gray = *Platanthera grandiflora* (Bigelow) Lindley
Habenaria strictissima Reichenbach *f.* var. *odontopetala* (Reichenbach *f.*) L. O. Williams = *Habenaria odontopetala* Reichenbach *f.*
Habenaria viridis Linnaeus var. *bracteata* (Mühlenberg *ex* Willdenow) Reichenbach *ex* Gray = *Coeloglossum viride* (Linnaeus) Hartman var. *virescens* (Mühlenberg) Luer

Habenaria viridis Linnaeus var. *interjecta* Fernald = *Coeloglossum viride* (Linnaeus) Hartman var. *virescens* (Mühlenberg) Luer
Habenella odontopetala Reichenbach *f.* (Small) = *Habenaria odontopetala* Reichenbach *f.*
Ibidium beckii (Lindley) House = *Spiranthes lacera* Rafinesque var. *gracilis* (Bigelow) Luer
Ibidium beckii House = *Spiranthes tuberosa* Rafinesque
Ibidium cernuum (Linnaeus) House = *Spiranthes cernua* (Linnaeus) L.C. Richard
Ibidium floridanum Wherry = *Spiranthes floridana* (Wherry) Cory *emend.* P. M. Brown
Ibidium gracile (Bigelow) House = *Spiranthes lacera* (Rafinesque) Rafinesque var. *gracilis* (Bigelow) Luer
Ibidium laciniatum (Small) House = *Spiranthes laciniata* (Small) Ames
Ibidium longilabre (Lindley) House = *Spiranthes longilabris* Lindley
Ibidium lucayanum Britton = *Mesadenus lucayanus* (Britton) Schlechter
Ibidium odoratum (Nuttall) House = *Spiranthes odorata* (Nuttall) Lindley
Ibidium ovale (Lindley) House = *Spiranthes ovalis* Lindley
Ibidium plantagineum (Rafinesque) House = *Spiranthes lucida* (H.H. Eaton) Ames
Ibidium praecox (Walter) House = *Spiranthes praecox* (Walter) S. Watson
Ibidium vernale (Englemann & A. Gray) House = *Spiranthes vernalis* Englemann & A. Gray
Isotria affinis (Austin *ex* A. Gray) Rydberg = *Isotria medeoloides* (Pursh) Rafinesque
Listera reniformis Small = *Listera smallii* Wiegand
Malaxis floridana (Chapman) O. Kuntze = *Malaxis spicata* Swartz
Malaxis unifolia Michaux var. *bayardii* nom. nud. = *Malaxis bayardii* Fernald
Mesadenus polyanthus (Reichenbach *f.*) Schlechter ≠ *Mesadenus lucayanus* (Britton) Schlechter
Microstylis unifolia (Michaux) Britton, Sterns, & Poggenberg = *Malaxis unifolia* Michaux
Neottia australis (Lindley) Szlachetko = *Listera australis* Lindley
Neottia gracilis Bigelow = *Spiranthes lacera* Rafinesque var. *gracilis* (Bigelow) Luer
Neottia smallii (Wiegand) Szlachetko = *Listera smallii* Wiegand
Orchis spectabilis Linnaeus = *Galearis spectabilis* (Linnaeus) Rafinesque
Physurus querceticola Lindley = *Platythelys querceticola* (Lindley) Garay
Platanthera blephariglottis (Willdenow) Lindley var. *conspicua* (Nash) Luer = *Platanthera conspicua* (Nash) P.M. Brown
Platanthera clavellata (Michaux) Luer = *Gymnadeniopsis clavellata* (Michaux) Rydberg *ex* Britton
Platanthera integra (Nuttall) Luer = *Gymnadeniopsis integra* Nuttall
Platanthera nivea (Nuttall) Luer = *Gymnadeniopsis nivea* Nuttall

Platanthera repens (Nuttall) Wood = *Habenaria repens* Nuttall
Platanthera ×vossii Case = ×*Platanthopsis vossii* (Case) P.M. Brown
Pogonia affinis Austin *ex* A. Gray = *Isotria medeoloides* (Pursh) Rafinesque
Pogonia bifaria P. M. Brown & R. P. Wunderlin = *Cleistes bifaria* (Fernald) Catling & Gregg
Pogonia divaricata (Linnaeus) R. Brown = *Cleistes divaricata* (Linnaeus) Ames
Pogonia pendula (Mühlenberg *ex* Willdenow) Lindley = *Triphora trianthophora* (Swartz) Rydberg
Pogonia verticillata (Mühlenberg *ex* Willdenow) Nuttall = *Isotria verticillata* (Mühlenberg *ex* Willdenow) Rafinesque
Spiranthes ×steigeri Correll = *Spiranthes ochroleuca* (Rydberg) Rydberg
Spiranthes beckii Lindley = *Spiranthes lacera* (Rafinesque) Rafinesque var. *gracilis* (Bigelow) Luer
Spiranthes brevilabris Lindley var. *floridana* (Wherry) Luer = *Spiranthes floridana* (Wherry) Cory
Spiranthes cernua (Linnaeus) L. C. Richard var. *ochroleuca* (Rydberg) Ames = *Spiranthes ochroleuca* (Rydberg) Rydberg
Spiranthes cernua (Linnaeus) Richard var. *odorata* (Nuttall) Correll = *Spiranthes odorata* (Nuttall) Lindley
Spiranthes gracilis (Bigelow) Beck = *Spiranthes lacera* (Rafinesque) Rafinesque var. *gracilis* (Bigelow) Luer
Spiranthes gracilis (Bigelow) L. C. Beck var. *brevilabris* (Lindley) Correll = *Spiranthes brevilabris* Lindley
Spiranthes gracilis (Bigelow) L. C. Beck var. *floridana* (Wherry) Correll = *Spiranthes floridana* (Wherry) Cory *emend.* P. M. Brown
Spiranthes grayi Ames = *Spiranthes tuberosa* Rafinesque
Spiranthes lanceolata (Aublet) Leon = *Sacoila lanceolata* (Aublet) Garay var. *lanceolata*
Spiranthes lucayana (Britton) Cogniaux = *Mesadenus lucayanus* (Britton) Schlechter
Spiranthes orchioides (Swartz) A. Richard = *Sacoila lanceolata* (Aublet) Garay
Spiranthes plantaginea Rafinesque = *Spiranthes lucida* (H.H. Eaton) Ames
Spiranthes polyantha Reichenbach ≠ *Mesadenus lucayanus* (Britton) Schlecter
Spiranthes simplex A. Gray = *Spiranthes tuberosa* Rafinesque
Spiranthes tuberosa var. *grayi* (Ames) Fernald = *Spiranthes tuberosa* Rafinesque
Stenorrhynchos lanceolatum (Aublet) Richard *ex* Sprengel = *Sacoila lanceolata* (Aublet) Garay var. *lanceolata*
Stenorrhynchos orchoides (Swartz) L.C. Richard = *Sacoila lanceolata* (Aublet) Garay
Tipularia unifolia Britton, Sterns, & Poggenberg = *Tipularia discolor* (Pursh) Nuttall
Triphora pendula (Mühlenberg *ex* Willdenow) Nuttall = *Triphora trianthophora* (Swartz) Rydberg
Triphora yucatanensis Ames ≠ *Triphora rickettii* Luer

Using Luer

Additions, corrections, nomenclatural changes, and comments for Luer (1972), *The Native Orchids of Florida,* and Luer (1975), *The Native Orchids of the United States and Canada excluding Florida* as pertaining to *Wild Orchids of the Southeastern United States.*

For those fortunate enough to own or have access to a copy of Carlyle Luer's original works on the orchids of Florida and the orchids of the United States and Canada, the following additions, corrections, and comments are assembled. These in no way should detract from the usefulness of those books, but simply allow for more than twenty-five years of research and nomenclatural changes as well as for the addition of several species that had not been described as of the dates of Luer's publications. Names of authors may be found in the text and checklist. No attempt has been made to completely rework the keys or the index.

The Native Orchids of Florida (1972)

Preface

pp. 8, 9, pl. 1:4 for *Cypripedium calceolus* var. *pubescens* read *Cypripedium parviflorum* var. *pubescens*

Introduction

Key
p. 29 couplet 31 contains all of the *Spiranthes* segregate genera
p. 30 couplet 41a for *Eulophia* read *Eulophia, Pteroglossaspis*
p. 31 couplet 52 for *Erythrodes* read *Platythelys*

Text

pp. 38, 39 pl. 6:5 forma *albiflora*

pp. 42, 43, pl. 8:6 forma *leucantha*

pp. 58, 59, pl. 15:8 *albiflora*

pp. 62, 63, pl. 16:1,6 forma *albiflorus*

pp. 66, 67, pl. 17:6 forma *viridis*

p. 90 couplet 8a for *Spiranthes grayi* read *Spiranthes tuberosa*

p. 90 couplet 10 for *Spiranthes brevilabris* var. *brevilabris* read *Spiranthes brevilabris*

p. 90 couplet 10a for *Spiranthes brevilabris* var. *floridana* read *Spiranthes floridana*

p. 90 couplet 12a for *Spiranthes cernua* var. *odorata* read *Spiranthes odorata*

p. 91 couplet 15a for *S. polyantha* read *Mesadenus lucayanus*

p. 91 couplet 19 for *S. lanceolata* var. *lanceolata* read *Sacoila lanceolata* var. *lanceolata*

p. 91 couplet 19a for *S. lanceolata* var. *luteoalba* read *Sacoila lanceolata* forma *albidaviridis*

pp. 100, 101, pl. 26:6–9 for *Spiranthes gracilis* read *Spiranthes lacera* var. *gracilis* none of the examples is from Florida; this species is not present in Florida

pp. 102, 103, pl. 27:1–6 for *Spiranthes brevilabris* var. *brevilabris* read *Spiranthes brevilabris*

pp. 102, 104, pl. 28:7–9 for *Spiranthes brevilabris* var. *floridana* read *Spiranthes floridana*

pp. 105, 106, pl. 28:1–5 for *Spiranthes grayi* read *Spiranthes tuberosa*

p. 109, pl. 29:2–5 are *Spiranthes odorata;* 6–8 for *Spiranthes cernua* var. *odorata* read *Spiranthes odorata*

p. 110 for *Spiranthes cernua* var. *odorata* read *Spiranthes odorata*

pp. 115, 116, pl. 31:5–8 for *Spiranthes polyantha* read *Mesadenus lucayanus*

pp. 117, 118, 119, pl. 32 for *Spiranthes lanceolata* var. *lanceolata* read *Sacoila lanceolata* var. *lanceolata*

pp. 120, 121, pl. 33 for *Spiranthes lanceolata* var. *luteoalba* read *Sacoila lanceolata* forma *albidaviridis*

p. 121, pl. 33:1–3 read *Sacoila lanceolata;* 4–6 read *Sacoila lanceolata* forma *albidaviridis*

p. 126, pl. 35:1,2,4,6 for *Erythrodes querceticola* read *Platythelys querceticola*

pp. 142, 143, pl. 40:3 for *P.* ×*chapmanii* read *P. chapmanii;* pl. 40:7 for *P.* ×*canbyi* read *P.* ×*channellii*

p. 151 for Platanthera ×*chapmanii* read *Platanthera chapmanii* and delete the hybrid combination of *P. ciliaris* × *P. cristata*

pp. 174, 175, pl. 51 read forma *albolabia*
p. 208 for *Epidendrum conopseum* read *Epidendrum magnoliae*
p. 236, couplet 1a for *Eulophia ecristata* read *Pteroglossaspis ecristata*
p. 238, pl. 72:6 for albino form read forma *pallida*
p. 240 for *Eulophia ecristata* read *Pteroglossaspis ecristata*

Taxa that are found in the southeastern United States and not treated in *The Native Orchids of Florida*

Spiranthes eatonii
Spiranthes ovalis var. *erostellata*
Spiranthes sylvatica

Hybrids:

Platanthera ×apalachicola
Platanthera ×beckneri
Platanthera ×lueri
Platanthera ×osceola
Spiranthes ×folsomii
Spiranthes ×itchetuckneensis
Spiranthes ×meridionalis

The Native Orchids of the United States and Canada excluding Florida (1975)

Introduction

p. 12 for *Cypripedium calceolus* var. *pubescens* read *Cypripedium parviflorum* var. *pubescens*

Text

p. 39, couplets 7 and 8, see *Cypripedium* key in text on page 53
pp. 40, 41, pl. 1:3,4 forma albiflorum
pp. 42, 43, pl. 2:1 forma albiflorum
p. 44 for *Cypripedium calceolus* Linnaeus var. *pubescens* (Willdenow) Correll read *Cypripedium parviflorum* Salisbury var. *pubescens* (Willdenow) Knight
pp. 44, 45, pl. 3, p. 46, 47, pl. 4 for *Cypripedium calceolus* var. *pubescens* read *Cypripedium parviflorum* var. *pubescens*
p. 48 for *Cypripedium calceolus* Linnaeus var. *parviflorum* Salisbury read *Cypripedium parviflorum* Salisbury var. *makasin* (Farwell) Sheviak
p. 49 for *Cypripedium ×andrewsii* Fuller read *Cypripedium ×andrewsii* Fuller nm. *andrewsii;* for *Cypripedium × favillianum* Curtis read *C. ×andrewsii* Fuller nm. *favillianum* (Curtis) Boivin; for *C. calceolus* var. *parviflorum* read *Cypripedium parviflorum* var. *makasin;* for *C. × landonii* read *C. ×andrewsii* nm. *landonii*

pp. 50, 51, pl. 5:1, 2 for *Cypripedium calceolus* var. *parviflorum* read *Cypripedium parviflorum* var. *makasin;* pl. 5:3,4,5, for *Cypripedium calceolus* Linnaeus var. *planipetalum* (Fernald) Victorin & Rousseau read *C. parviflorum* var. *pubescens* extreme expression

pp. 52, 53, pl. 6:5 read *Cypripedium* ×*andrewsii* nm. *favillianum;* pl. 6:6 read *Cypripedium* ×*andrewsii* nm. *andrewsii*

pp. 56, 57, pl. 8:5 for (*Cypripedium album*) read forma *albiflorum*

pp. 76,77, pl 15:3 forma *viridens*

pp. 82, 83, pl. 17:5 forma *viridis*

pp. 84, 85, pl. 18:4 forma *viridens*

p. 99, couplet 14a for *S. brevilabris* var. *floridana* read *S. floridana*

p. 100, couplet 27a for *S. polyantha* read *Mesadenus lucayanus*

p. 101, couplet 34 for *S. lanceolata* read *Sacoila lanceolata;* couplet 34a for *S. lanceolata* var. *luteoalba* read *Sacoila lanceolata* var. *lanceolata* forma *albidaviridis*

p. 102 (chart) for *brevilabris* var. *brevilabris* read *brevilabris;* for *brevilabris* var. *floridana* read *floridana*

p. 105, pl. 22:5 forma *albolabia*

p. 114 read var. *ovalis*

pp. 144, 145, pl. 35:7 forma *ophioides*

pp. 148, 149, pl. 36:4 forma *willeyi;* 5 forma *gordinierii;* for *Cypripedium calceolus* var. *pubescens* read *Cypripedium parviflorum* var. *pubescens*

p. 177, couplet 9 for *P.* × *chapmanii* read *P. chapmanii*

p. 178, couplet 19 for *P. integra* read *Gymnadeniopsis integra;* couplet 19a for *P. nivea* read *Gymnadeniopsis nivea;* couplet 20a for *P. clavellata* read *Gymnadeniopsis clavellata*

pp. 184, 185 for *Platanthera blephariglottis* var. *conspicua* read *Platanthera conspicua*

p. 185:4 is *Platanthera* ×*lueri*

p. 187 for *P.* × *chapmanii* (Small) Luer read *P. chapmanii* (Small) Luer *emend.* Folsom

pp. 188, 189, pl. 46:4,6 for *P.* × *chapmanii* read *P. chapmanii;* 46:7 *P.* ×*beckneri*

pp. 196, 197, pl. 49:8 forma *albiflora*

pp. 202, 203 for *Platanthera integra* read *Gymnadeniopsis integra*

pp. 204, 205 for *Platanthera nivea* read *Gymnadeniopsis nivea*

pp. 206, 207 for *Platanthera clavellata* read *Gymnadeniopsis clavellata*

pp. 246, 247, pl. 66:3,5 forma *leucantha;* pl. 66:4 for var. *bifaria* read *Cleistes bifaria*

p. 248 for var. *bifaria* read *Cleistes bifaria*

pp. 260, 261, pl. 70:3 forma *albiflorus*

pp. 266, 267, pl. 73:4 forma *albiflora;* pl. 73:5 forma *subcaerulea*

pp. 270, 271, pl. 74:1 forma *albolabia*
pp. 320, 321, pl. 90:3 forma *albolabia*
p. 328:3, 329, pl. 94:3 is *Corallorhiza wisteriana* forma *albolabia*

Taxa that are found in the southeastern United States and not treated in *The Native Orchids of the United States and Canada excluding Florida* (1975)
Calopogon oklahomensis
Corallorhiza odontorhiza var. *pringlei*
Cypripedium kentuckiense
Cypripedium parviflorum var. *parviflorum*
Spiranthes eatonii
Spiranthes ovalis var. *erostellata*
Spiranthes sylvatica
Hybrids:
Platanthera ×*apalachicola*
Platanthera ×*beckneri*
Platanthera ×*channellii*
Platanthera ×*keenanii*
Platanthera ×*lueri*
Platanthera ×*osceola*
×*Platanthopsis vossii*
Spiranthes ×*folsomii*
Spiranthes ×*itchetuckneensis*
Spiranthes ×*intermedia*
Spiranthes ×*meridionalis*

Species Pairs

When two species appear to be very closely related and have a similar morphology they are often referred to as species pairs. Although not necessarily a scientific or taxonomic term, the designation is often helpful in recognizing two species that are so alike it is often difficult to determine one from the other whether in the field or in the herbaria. Their taxonomic history usually involves synonyms and/or recognition at different taxonomic levels, i.e., subspecies, varieties, or rarely forma. Validation of the two taxa at species level usually involves studies of pollinators, habit, habitat, range, morphology, and, in more recent years, DNA analyses. Six such species pairs are found in the Southeast. Fortunately a few simple morphological characters easily separate the taxa. In each case these are characters that can be found within the key to the species. Also, range and habitat are often well separated.

Calopogon barbatus and *Calopogon multiflorus*

Calopogon barbatus (Walter) Ames

bearded grass-pink

widespread throughout most of the Coastal Plain; petals widest below the middle

Calopogon multiflorus Lindley

SYNONYM *Calopogon barbatus* var. *multiflorus* (Lindley) Correll

many-flowered grass-pink

rare and local; petals widest above the middle; see descriptions on pages 26 and 28 for details

Cleistes bifaria and *Cleistes divaricata*

Cleistes bifaria (Fernald) Catling & Gregg

SYNONYM *Cleistes divaricata* var. *bifaria* Fernald

upland spreading pogonia

widespread throughout the Florida Panhandle; lip about 27 mm long, broadly pointed at the apex

Cleistes divaricata (Linnaeus) Ames

large spreading pogonia

very rare and local in northeastern Florida and the Atlantic Coastal Plain; lip 34–56 mm long, narrowly pointed at the apex; see descriptions on pages 000 and 000 for details

Malaxis bayardii and *Malaxis unifolia*

Malaxis bayardii Fernald

Bayard's adder's-mouth

Very rare in the east coastal states; found either in the sandy Coastal Plain areas or in the rocky uplands of the Appalachian foothills.

Lower flowers persistent after pollination, raceme broadly cylindrical, prominent auricles on the lip

Malaxis unifolia Michaux

green adder's-mouth

Lower flowers withering after pollination, inflorescence a flat-topped, broadened panicle, auricles not prominent on the lip, wide-ranging throughout the eastern half of North America

Because of the tiny flowers on these two species of *Malaxis,* the flower characters are difficult to see. The shape of the inflorescence and persistent flowers are a great aid.

See descriptions on pages 000 and 000 for details.

Platanthera blephariglottis and *Platanthera conspicua*

Platanthera blephariglottis (Willdenow) Lindley

northern white fringed orchis

eastern half of North America, terminating in scattered populations in South Carolina and Georgia; habitat is typically open meadows, bogs, wet power lines, sphagnous roadsides

Platanthera conspicua (Nash) P. M. Brown

southern white fringed orchis

SYNONYM *Platanthera blephariglottis* var. *conspicua*
restricted to the southeastern Coastal Plain from North Carolina west to eastern Texas; in overlapping areas populations are distinct in habitat and flowering time.

In only a few areas of South Carolina and southeastern Georgia should there be any confusion between these two distinct species. The much larger flowers, with a narrowed isthmus on the lip and the very long spur, make *Platanthera conspicua* distinctive. Large plants of *P. blephariglottis* from New Jersey southward have often been erroneously identified as *P. conspicua.*

See descriptions on pages 136 and 192 for details.

Platanthera grandiflora and *Platanthera psycodes*

Platanthera grandiflora (Bigelow) Lindley

SYNONYM *Habenaria psycodes* var. *grandiflora*
large purple fringed orchis

racemes 3–5 cm in diameter, spur orifice rounded, entire raceme is eventually opened simultaneously (lower flowers *not* withering as upper flowers open)

Platanthera psycodes (Linnaeus) Lindley

small purple fringed orchis

racemes 1.0–1.5 cm in diameter flowers opening sequentially, spur orifice a transverse dumbbell

Within the range of this book these two species occur only in northern Georgia, and then are very rare. Flowering times are usually separated by several weeks; see descriptions on pages 150–51 and 160–61 for details.

Spiranthes brevilabris and *Spiranthes floridana*

Spiranthes brevilabris Lindley

short-lipped ladies'-tresses

plants densely pubescent

Spiranthes floridana (Wherry) Cory *emend.* P. M. Brown

SYNONYM *Spiranthes brevilabris* var. *floridana* (Wherry) Luer
Florida ladies'-tresses

plants essentially glabrous; ranges and habitats overlap throughout much of the southeastern Coastal Plain; see descriptions on pages 192 and 198 for details

Part 4

Orchid Hunting Throughout the Southeast

Note: The seasons come at different months throughout the Southeast. For that reason months are rarely given in the following selections. Seasons are referred to and should be adjusted for the particular area you are exploring.

1. East Texas and Big Thicket

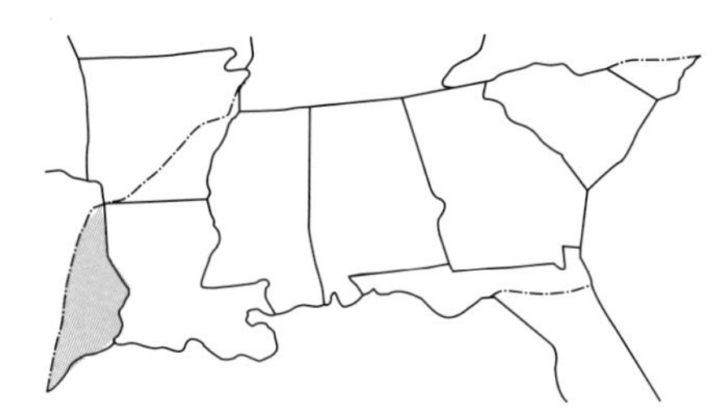

The Big Thicket area of eastern Texas once comprised nearly 200,000 acres. Today it covers a much smaller area, set aside as Big Thicket National Preserve. The boundaries of the historical Big Thicket have always been in dispute but the national preserve consists of twelve units of different sizes and locales totaling 84,500 acres. The preserve located near Beaumont, Texas, east of I-45, is bordered on the east by Louisiana. Although many will lament that in the past the area contained endless numbers of wildflowers, including orchids, there are still many species to be found. Roads may be few but exploration is unlimited. All of Texas east of I-45 presents a larger array of species than Big Thicket. This area includes the Angelina National Forest, home to several species not found elsewhere in eastern Texas.

Species typically found in eastern Texas and Big Thicket:

Calopogon oklahomensis, Oklahoma grass-pink
Calopogon tuberosus, common grass-pink
Corallorhiza odontorhiza var. *odontorhiza,* autumn coralroot
Corallorhiza wisteriana, Wister's coralroot
Cypripedium kentuckiense, ivory-lipped lady's-slipper
Gymnadeniopsis clavellata, little club-spur orchis
Gymnadeniopsis integra, yellow fringeless orchis
Gymnadeniopsis nivea, snowy orchis
Hexalectris spicata, crested coralroot
Isotria verticillata, large whorled pogonia

Listera australis, southern twayblade
Malaxis unifolia, green adder's-mouth
Platanthera chapmanii, Chapman's fringed orchis
Platanthera ciliaris, orange fringed orchis
Platanthera cristata, orange crested orchis
Platanthera flava var. *flava,* southern tubercled orchis
Platanthera lacera, green fringed orchis, ragged orchis
Pogonia ophioglossoides, rose pogonia
Ponthieva racemosa, shadow-witch
Spiranthes brevilabris, short-lipped ladies'-tresses
Spiranthes cernua, nodding ladies'-tresses
Spiranthes eatonii, Eaton's ladies'-tresses
Spiranthes floridana, Florida ladies'-tresses
Spiranthes lacera var. *gracilis,* southern slender ladies'-tresses
Spiranthes odorata, fragrant ladies'-tresses
Spiranthes ovalis var. *ovalis,* southern oval ladies'-tresses
Spiranthes parksii, Navasota ladies'-tresses
Spiranthes praecox, giant ladies'-tresses
Spiranthes sylvatica, woodland ladies'-tresses
Spiranthes tuberosa, little ladies'-tresses
Spiranthes vernalis, grass-leaved ladies'-tresses
Tipularia discolor, crane-fly orchis
Triphora trianthophora, three birds orchid

Spring in eastern Texas may not produce the orchid bonanza of many other areas, but it still provides us with an abundance of general wildflowers and a respectable show of native orchids. As in many other areas of the Southeast, spring starts with *Corallorhiza wisteriana,* Wister's coralroot, soon to be followed by *Listera australis,* the southern twayblade. *Isotria verticillata,* the large whorled pogonia and, although a bit rarer, *Malaxis unifolia,* the green adder's-mouth, also come into early spring flower. All four of these species are found in mixed woodlands, although the *Listera* certainly prefers a moister area than the others.

The open prairies and wetlands will then be painted in various shades of pink and white with *Calopogon oklahomensis,* the Oklahoma grass-pink, soon to be followed by the more frequently encountered *Calopogon tuberosus,* the common grass-pink. Plants from eastern Texas previously identified as *Calopogon barbatus,* the bearded grass-pink, have all proven to be the recently described Oklahoma grass-pink.

By the time spring is fully underway the ladies'-tresses start to appear along the roadways and even in the woodlands. *Spiranthes vernalis,* the grass-leaved ladies'-tresses, is the most frequently seen member of the genus and is found throughout the entire Southeast. Less abundant, but still well distributed are *S. praecox,* the giant ladies'-tresses, and *S. sylvatica,* the woodland ladies'-tresses. The latter has recently been described and has the characteristic of preferring shaded woodlands and of being tucked up under the outer edge of thickets. Both species have the distinctive green veining in the lip. Three additional ladies'-tresses also flower at this time and all are apparently very rare and their status uncertain. *Spiranthes brevilabris,* the short-lipped or Texas ladies'-tresses, *S. floridana,* the Florida ladies'-tresses, and *S. eatonii,* Eaton's ladies'-tresses, all have similar morphologies but distinctive characters that separate them. Read the three species descriptions carefully, for despite their extreme rarity, any or all may still be found in East Texas.

The absolute gem of the late spring woodlands is the rare *Cypripedium kentuckiense,* the ivory-lipped lady's-slipper. The largest-flowered of any of our yellow lady's-slippers, this spectacular species was once a common component of the spring woodland flora of western Arkansas, northwestern Louisiana, and eastern Texas. Development and logging have greatly reduced the number of locations for this choice orchid, but it can still be seen in several places both within and apart from Big Thicket.

Early summer is certainly the beginning of the high season for the many fringed orchises and their cousins. This group runs the gamut from the very rarest to the most common. The first to show itself is *Gymnadeniopsis nivea,* the snowy orchis. Often favoring broad, mowed roadsides and damp grassy savannas, these small spikes of stark white flowers stand out among the surrounding vegetation. A short foray to nearby woodlands may be rewarded with a pleasant surprise of *Hexalectris spicata,* the crested coralroot—a surprise only because it is unpredictable in its flowering from year to year. Back in the open roadside ditches and swamps, stands of *Platanthera flava* var. *flava,* the southern tubercled orchis, may sometimes be found in colonies of more than a hundred plants.

Platanthera chapmanii, Chapman's fringed orchis, *P. ciliaris,* the orange fringed orchis, and *P. cristata,* the orange crested orchis, make up a colorful midsummer trio that is well distributed throughout the Big Thicket. The rarest of these, Chapman's fringed orchis, is one of the most geographically restricted orchids we have in North America. Apart from the central Panhandle of Florida and a historic site in southeastern Georgia, the only other known area for this species is eastern Texas, primarily in Big Thicket. The other two orange orchises are found in a much wider area.

Farther afield from the Big Thicket area, but still in East Texas, is the lone historic site for *Platanthera lacera,* the green fringed or ragged orchis. It was first recorded from here in 1946 and has not been seen since in Texas, but elsewhere to the northeast it can be a common species. *Gymnadeniopsis integra,* the yellow fringeless orchis, holds a similar record, as it was not found in Texas until 1987, in the Angelina National Forest. Prior to that it had been documented but once, and that documentation was mislabeled as *G. nivea,* whereas an earlier collection labeled *G. integra* actually was *G. nivea*!

Concluding summer are two woodland species, both scattered and rare, but posing no problems of identification. *Tipularia discolor,* the crane-fly orchis, and *Triphora trianthophora,* the three birds orchid, are both reaching the western limit of their ranges in eastern Texas, and both occupy similar niches in the rich woodlands. With *Tipularia* the more frequently seen of the two, *Triphora* is not only rarer but much harder to locate in flower. The flowering period is for only a few hours in each day and then again for only a few days. It takes both diligence and patience to capture the three birds at their best!

Autumn is the time for *Corallorhiza odontorhiza* var. *odontorhiza,* the autumn coralroot, and *Ponthieva racemosa,* the shadow-witch. The coralroot is a slender and obscure denizen of deciduous woodlands, and often flowers under fallen leaves with its tiny cleistogamous blooms. The shadow-witch prefers shaded, moist river floodplains and slopes.

Autumn is also the time for more ladies'-tresses. *Spiranthes lacera* var. *gracilis,* the southern slender ladies'-tresses, *S. cernua,* the nodding ladies'-tresses, and *S. odorata,* the fragrant ladies'-tresses, are all frequently seen species in this genus. The southern slender ladies'-tresses usually grows in dry short-grass fields, lawns, and old cemeteries, whereas the fragrant ladies'-tresses is a wetland species and may be found in both sun and shade, often in shallow standing water. *Spiranthes cernua* is the most widespread and frequently seen ladies'-tresses in North America and occurs in many regional races. Here in eastern Texas one of the most unusual races is a cleistogamous/peloric race with yellow or green unopened flowers. Typically *S. cernua* has crystalline, nodding white to ivory flowers, often with the fragrance of vanilla.

Our yearlong tour of orchids in eastern Texas concludes with the federally endangered *Spiranthes parksii,* the Navasota ladies'-tresses. Known primarily from the Post Oak Savannah region near College Park, Texas, this very unusual species was discovered in 1945, declared extinct in 1975, and rediscovered in 1978 within the same Post Oak Savannah area and not more than ten miles from where it was originally found. In October 1986 the species was found for the first time in Angelina National Forest in the Pineywoods region of Jasper County, more than a hundred miles east of the original sites. *Spiranthes parksii,* with its partially open, off-white flowers, certainly is not the most attractive of our ladies'-tresses, but does demand a pilgrimage for those enslaved by the genus!

2. Gulf Coastal Plain

Extending from the Big Bend of western Florida to along the northeastern coast of Texas, the Gulf Coastal Plain provides habitat for the majority of species of orchids found in the southeastern United States. One needs to understand a bit about the geography and hydrology of the Southeast to put this habitat into perspective. For the purposes of this book the Gulf Coastal Plain excludes the Mississippi Delta and Bayou regions found primarily in Louisiana since they are treated on page 336. But the region also includes much of southern Arkansas, defined as the West Gulf Coastal Plain region. The sandy soils are all similar in these areas and subsequently the distribution of orchid species is also similar.

Interspersed within this coastal plain area are local hardwood hammocks, longleaf pine flatwoods, and agriculturally disturbed areas, including timber harvests and rice fields. All of these habitats have their own orchid specialties and it does not take a great deal of exploring to find a common thread. Although no species reaches its eastern limit here, many reach their western boundaries, and a few are local disjuncts. Most of the species listed below are more frequent eastward and rare and local as they progress west, and in a few cases absent or historical in Texas. Those species marked # are more typical of upland forests, though they do rarely occur on the Gulf Coastal Plain as well.

Species typically found on the Gulf Coastal Plain:

Calopogon barbatus, bearded grass-pink
Calopogon multiflorus, many-flowered grass-pink

Calopogon pallidus, pale grass-pink
Calopogon tuberosus, common grass-pink
Cleistes bifaria, upland spreading pogonia
Corallorhiza wisteriana, Wister's coralroot
Epidendrum magnoliae, green-fly orchis
Gymnadeniopsis integra, yellow fringeless orchis
Gymnadeniopsis nivea, snowy orchis
Habenaria quinqueseta, Michaux's orchid
Habenaria repens, water spider orchid
Isotria verticillata, large whorled pogonia #
Listera australis, southern twayblade
Malaxis spicata, Florida adder's-mouth
Malaxis unifolia, green adder's-mouth #
Platanthera chapmanii, Chapman's fringed orchis
Platanthera ciliaris, orange fringed orchis
Platanthera conspicua, southern white fringed orchis
Platanthera cristata, orange crested orchis
Platanthera flava var. *flava*, southern tubercled orchis
Platythelys querceticola, low ground orchid, jug orchid
Pogonia ophioglossoides, rose pogonia
Ponthieva racemosa, shadow-witch
Pteroglossaspis ecristata, crestless plume orchid
Spiranthes brevilabris, short-lipped ladies'-tresses
Spiranthes cernua, nodding ladies'-tresses
Spiranthes eatonii, Eaton's ladies'-tresses
Spiranthes floridana, Florida ladies'-tresses
Spiranthes laciniata, lace-lipped ladies'-tresses
Spiranthes longilabris, long-lipped ladies'-tresses
Spiranthes odorata, fragrant ladies'-tresses
Spiranthes ovalis, oval ladies'-tresses #
Spiranthes praecox, giant ladies'-tresses
Spiranthes sylvatica, woodland ladies'-tresses
Spiranthes tuberosa, little ladies'-tresses
Spiranthes vernalis, grass-leaved ladies'-tresses
Tipularia discolor, crane-fly orchis #
Triphora trianthophora, three birds orchid #
Zeuxine strateumatica, lawn orchid*

Of the four grass-pinks, only *Calopogon tuberosus*, the common grass-pink, is frequent throughout the entire area. *Calopogon pallidus*, the pale grass-pink, and

C. barbatus, the bearded grass-pink, both become increasingly rare in Louisiana and are absent from Texas. *Calopogon multiflorus,* the many-flowered grass-pink, reaches the western limit of its range in eastern Louisiana as well. Several of the records for *C. barbatus* in Louisiana and eastern Texas have proven to be *C. oklahomensis,* with its habitat defined more appropriately as prairie than coastal plain.

Cleistes bifaria, the upland spreading pogonia, extends sparingly westward and is known from only a few locations in Mississippi, Louisiana, and (historically) Texas. Both *Corallorhiza wisteriana,* Wister's coralroot, and *Epidendrum magnoliae,* the green-fly orchis, are widespread along the Gulf Coastal Plain and the green-fly orchis, our only epiphytic orchid north of Florida, extends to a few miles east of the Texas state line, but has never been confirmed for Texas. Wister's coralroot is a common species not only here but northward as well. *Listera australis,* the southern twayblade, is often found in wet woodlands of the same areas.

Only two species of true rein orchises are known from this region. The first, *Habenaria repens,* the water spider orchid, is the only aquatic orchid found in the United States and is well distributed throughout the Southeast, while *Habenaria quinqueseta,* Michaux's orchid, although found regularly in the Panhandle of Florida, is very rare elsewhere and historical in Texas.

The open wet grasslands, pine flatwoods, and—not surprisingly—damp, mown roadsides provide habitat for the delightful *Pogonia ophioglossoides,* the rose pogonia, in the spring and several months later, for a whole set of showy summer-flowering species. Although *Platanthera chapmanii,* Chapman's fringed orchis, is very restricted in its distribution to the Panhandle of Florida and eastern Texas on the Gulf Coastal Plain, it is often found growing in the same areas as *Platanthera ciliaris,* the orange fringed orchis, *Platanthera conspicua,* the southern white fringed orchis, and *Platanthera cristata,* the orange crested orchis. These three closely related orange-flowered species can easily form hybrid swarms, which in turn can make identification more difficult. Colonies in the Apalachicola and Osceola National Forests, while certainly on or near the Gulf Coastal Plain, are classic for these hybrid swarms, which are treated on pages 168–71. The southern white fringed orchis currently ranges westward only to southern Louisiana and is known from Texas only as a historical record. Two additional species, formerly treated in the genus *Platanthera, Gymnadeniopsis integra,* the yellow fringeless orchis, and *Gymnadeniopsis nivea,* the snowy orchis, both occupy habitats in this same region and range northward as well. *Gymnadeniopsis integra* was only recently rediscovered in Texas (1988) and does not range into southern Arkansas.

The largest genus in the southeastern United States is *Spiranthes,* the ladies'-tresses. With fifteen species in the region there is no area that does not abound with both plants and species. They are most prolific along the Gulf Coastal Plain, with both common and rare species well distributed.

Spiranthes brevilabris, the short-lipped ladies'-tresses, *Spiranthes eatonii,* Eaton's ladies'-tresses, and *Spiranthes floridana,* Florida ladies'-tresses, form a spring-flowering trio very similar morphologically and often mistaken for each other. Many collections and currently identified sites for both *Spiranthes brevilabris* and *S. floridana* have proven to be *S. eatonii,* a new species described in 1999. The two former species are exceedingly rare and known currently from only one or two confirmed locations.

A second spring-flowering trio, comprising *Spiranthes praecox,* the giant ladies'-tresses, *S. sylvatica,* the woodland ladies'-tresses, and *S. vernalis,* the grass-leaved ladies'-tresses, is widely distributed throughout the Southeast and relatively common. However, recent investigations raise a question as to whether *S. praecox* is present in Arkansas; all plants previously identified as *S. praecox* appear to be *S. sylvatica,* another recently described species (2001).

As early summer commences two additional ladies'-tresses are found distributed throughout the Gulf Coastal Plain and well beyond. They are *Spiranthes laciniata,* the lace-lipped ladies'-tresses—a species of very wet habitats—and *S. tuberosa,* the little ladies'-tresses—a species of very dry habitats. Autumn brings another three species, one fairly common, *S. odorata,* the fragrant ladies'-tresses, one much rarer—*S. longilabris,* the long-lipped ladies'-tresses—and the third a great rarity—*S. ovalis,* the oval ladies'-tresses. The first two are often found in similar damp and wet roadsides and open flatwoods and, not surprisingly, the hybrid of the two, *S. ×folsomii,* can be present. Curiously enough, the latter two species can also be found in a shared habitat of wet, rich woodlands and also produce a hybrid—*S. ×itchetuckneensis.*

These same rich, moist woodlands can, in the very early spring, also harbor *Listera australis,* the southern twayblade, while later in the season, *Malaxis spicata,* the Florida adder's-mouth, *Platythelys querceticola,* the low ground orchid, and *Platanthera flava* var. *flava,* the southern tubercled orchis, may be present. This combination can be found with some ease in the eastern Panhandle of Florida; as one progresses westward the *Malaxis* and *Platythelys* cease to occur and only in central Louisiana does the *Platythelys* reappear. It should be carefully sought along the coastal hammocks of Alabama and Mississippi.

Scattered throughout many of the hardwood hammocks and open swamps can be found plants of *Ponthieva racemosa,* the shadow-witch, *Tipularia discolor,* crane-fly orchis, *Triphora trianthophora,* the three birds orchid, and, in more open areas, *Pteroglossaspis ecristata,* the crestless plume orchid. The latter two are increasingly rare and *Pteroglossaspis* does not occur in Texas.

Two other species not regularly typical of the Gulf Coastal Plain region that could also be found are *Isotria verticillata,* the large whorled pogonia, and *Malaxis unifolia,* the green adder's-mouth, both species more typical of the Atlantic Coastal Plain.

Finally two non-native species have been found in this area. *Zeuxine strateumatica,* the lawn orchid, which is thoroughly naturalized throughout Florida and perhaps, through the distribution of nursery stock, has found its way as far west as Texas, and *Bletilla striata,* the urn orchid, a presumed garden escape known only from Escambia County, Florida.

3. Apalachicola and Osceola National Forests; Greater Jacksonville

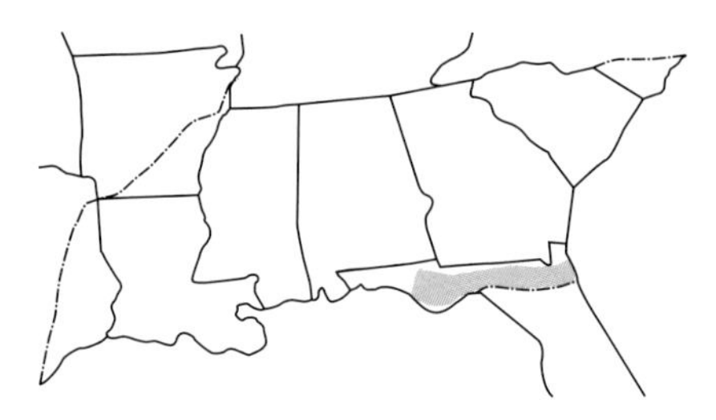

The National Forests

Nestled within the Panhandle and northern counties of Florida, and well within the Coastal Plain, are two of the larger national forests in the eastern United States. Although at first glance they may appear to be much the same, they are really very different from one another. The Apalachicola National Forest, situated along the Apalachicola River, is a mix of pine flatwoods and broad savannas riddled with numerous streams draining into the Gulf of Mexico. The Osceola National Forest is higher and drier and, even though it has its share of wetlands, presents a much more forested habitat. Although the plant list is more limited for these two sites than for some other places, these areas are nonetheless very specialized and always worth the time it takes to explore them.

Species typical of the Apalachicola and Osceola National Forests (A = Apalachicola National Forest only; O = Osceola National Forest only):

Calopogon barbatus, bearded grass-pink
Calopogon multiflorus, many-flowered grass-pink
Calopogon pallidus, pale grass-pink
Calopogon tuberosus, common grass-pink
Cleistes bifaria, upland spreading pogonia
Cleistes divaricata, spreading pogonia O

Corallorhiza wisteriana, Wister's coralroot
Gymnadeniopsis integra, yellow fringeless orchis A
Gymnadeniopsis nivea, snowy orchis
Habenaria quinqueseta, Michaux's orchid
Habenaria repens, water spider orchid
Listera australis, southern twayblade
Platanthera chapmanii, Chapman's fringed orchis
Platanthera ciliaris, orange fringed orchis O
Platanthera conspicua, southern white fringed orchis O
Platanthera cristata, orange crested orchis
Platanthera flava var. *flava,* southern tubercled orchis
Platythelys querceticola, low ground orchid, jug orchid A
Pogonia ophioglossoides, rose pogonia
Ponthieva racemosa, shadow-witch
Pteroglossaspis ecristata, crestless plume orchid
Spiranthes brevilabris, short-lipped ladies'-tresses O
Spiranthes cernua, nodding ladies'-tresses A
Spiranthes eatonii, Eaton's ladies'-tresses
Spiranthes floridana, Florida ladies'-tresses
Spiranthes laciniata, lace-lipped ladies'-tresses
Spiranthes longilabris, long-lipped ladies'-tresses A
Spiranthes odorata, fragrant ladies'-tresses
Spiranthes ovalis var. *ovalis,* southern oval ladies'-tresses A
Spiranthes ovalis var. *erostellata,* northern oval ladies'-tresses A
Tipularia discolor, crane-fly orchis

Early spring brings a similar list to both forests, with possibly all four of the grass-pinks in suitable habitat. The earliest is *Calopogon barbatus,* the bearded grass-pink, which is widespread and often found along damp roadside bankings and ditches. It is soon followed by the rarest of the quartet, *C. multiflorus,* the many-flowered grass-pink. Although never a guaranteed species, careful searching in areas that have been burned four to six weeks earlier may reveal a few flowering plants. Soon after this, usually by mid-April, *C. tuberosus,* the common grass-pink, is abundant in both forests, followed by the unusual *C. pallidus,* the pale grass-pink, with its upswept lateral petals. Often accompanying many of the grass-pinks is the delightful little *Pogonia ophioglossoides,* rose pogonia, frequently growing in great masses in the open savannas.

This is also the time to start searching for *Cleistes bifaria,* the upland spreading pogonia. This species has a preference for recently burned areas as well, and will flower for several years following a burn. As the woodland regrows the plants then produce only leaves until another burn. Only in the Osceola National Forest in wet woods along CR 210 can one find *C. divaricata,* the spreading pogonia. Here it is found in a bordering woodland growing with *Platanthera cristata.*

Spring also brings a variety of ladies'-tresses as well, the most noticeable being *Spiranthes vernalis,* the grass-leaved ladies'-tresses, *S. praecox,* the giant ladies'-tresses, and *S. sylvatica,* the woodland ladies'-tresses. There are also records in both national forests for the rare spring trio of *S. brevilabris,* short-lipped ladies'-tresses, *S. eatonii,* Eaton's ladies'-tresses, and *S. floridana,* the Florida ladies'-tresses. Eaton's ladies'-tresses is the most likely to be found in either forest, as it prefers open, dry, sandy areas. The other two species are exceedingly rare and should be carefully sought.

Gymnadeniopsis nivea, the snowy orchis, is the first of the summer specialties to flower, particularly along Florida Route 65 in the Sumatra area. This area is home to many very rare and endangered species, not only orchids. The county, as well as the Forest Service, manages this area to encourage and protect these rare plants. The broad, open, wet savannas that support the snowy orchis later in the summer may reveal stands of *G. integra,* the yellow fringeless orchis, one of the rarest orchids in the Southeast. The open pine flatwoods in both national forests support large mixed colonies of *Platanthera chapmanii,* Chapman's fringed orchis, and *Platanthera cristata,* the orange crested orchis. The hybrid of these, P. ×*apalachicola,* is usually present as well.

Chapman's fringed orchis is endemic to the Southeast and found only in northern Florida, one place (historically) in nearby Georgia, and East Texas. *Platanthera ciliaris,* the orange fringed orchis, is also present in Osceola National Forest along with the hybrid *P.* ×*osceola.* Hybrid swarms develop in both forests.

In addition *Platanthera conspicua,* the southern white fringed orchis, is found in the Osceola National Forest along Florida Route 90. *Platanthera* ×*beckneri,* Beckner's hybrid fringed orchis, *P.* ×*channellii,* Channell's hybrid fringed orchis, *P.* ×*lueri,* Luer's hybrid fringed orchis, and occasional backcrosses also occur. It can be a real challenge to try and identify, let alone sort out, all of these various combinations. Details are given in the various species accounts to assist in this task.

Late summer and autumn bring primarily more species of ladies'-tresses, but also *Pteroglossaspis ecristata,* the crestless plume orchid, *Habenaria quinqueseta,* Michaux's orchid, and *H. repens,* the water spider orchid (which may flower at any time of year); in the shaded swamps, *Platythelys querceticola,* the low ground orchid or jug orchid, and the delicate *Ponthieva racemosa,* the shadow-witch orchid, can be found. Earlier in the year many of these same swamps may very well have had *Platanthera flava* var. *flava,* the southern tubercled orchis.

As autumn progresses many of the wet roadside seeps and ditches will come alive with *Spiranthes odorata,* the fragrant ladies'-tresses, also equally at home in the wooded swamps. Drier areas in those swamps and adjacent damp woodlands may also reveal the very rare *S. ovalis* var. *ovalis,* southern oval ladies'-tresses, and/or *S. ovalis* var. *erostellata,* the northern oval ladies'-tresses. The latter variety is known from only a few sites in Florida. Equally as rare, but a bit easier to search for, is *S. longilabris,* the long-lipped ladies'-tresses, inhabiting damp road shoulders, pine flatwoods, and open savannas, often with the fragrant ladies'-tresses. Should you find these two growing together you should carefully search for *S.* ×*folsomii,* their hybrid.

The rarest of the ladies'-tresses to be found in northern Florida occurs along a few muddied stream runoffs within the Apalachicola National Forest. It is *Spiranthes cernua,* the nodding ladies'-tresses. *Spiranthes cernua* is one of the most common orchids in the eastern United States, but this Deep South race of the species is known from only this area of the national forest. It flowers well into December and is the latest orchid species found in the calendar year.

In some of the deeper live oak hammocks that dot both of the national forests one should search for *Corallorhiza wisteriana,* Wister's coralroot, in February. Also present is *Epidendrum magnoliae,* the green-fly orchis. It is the only epiphytic species seen in the Southeast and may flower at most any time of year.

Continuing east of the national forests to the northeastern corner of Florida we come to Duval and Nassau counties, which comprise the greater Jacksonville area. This lush corner of Florida offers several species not otherwise seen in the Southeast.

Greater Jacksonville

Additional species found in the Jacksonville area:

Cleistes divaricata, spreading pogonia
Epidendrum magnoliae, green-fly orchis
Habenaria odontopetala, toothed rein orchis
Mesadenus lucayanus, copper ladies'-tresses

The verdant roadsides, meadows, and pine flatwoods along US 301 and A1A in northeastern Florida have many of the same species seen in both of the national forests. These unusually wide road shoulders that are occasionally mown become ideal habitat for several species of the ladies'-tresses, most notably *Spiranthes vernalis,* the grass-leaved ladies'-tresses. The dense live oak hammocks are full of *Epidendrum magnoliae,* the green-fly orchis, flowering at random times throughout the year. The pine flatwoods in very selected areas still harbor a few plants of *Cleistes divaricata,* the spreading pogonia, which is best seen in early May. In addition the Jacksonville area also has *Mesadenus lucayanus,* the copper ladies'-tresses, known locally as the Ft. George ladies'-tresses for its northernmost location at Ft. George State Park, and *Habenaria odontopetala,* the toothed rein orchis, another southern species that is also reaching the northern limit of its range within one of the many municipal parks in the City of Jacksonville.

Although not technically within the definition of the Southeast for this book, it is important to mention Jennings State Forest and the vicinity of Middleburg, Florida, in Clay County. *Platanthera chapmanii* and *Spiranthes eatonii* both grow here, and the state forest as well as the roadsides within Middleburg are certainly hot spots for a long list of wild orchids, including both species of *Cleistes.*

4. The Great Temperate Hammocks of the Southeast

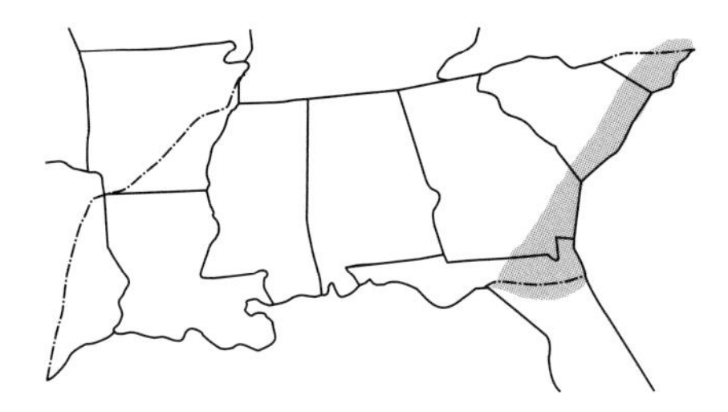

One of the major components of the Coastal Plain and adjacent Piedmont of the southeastern United States is the ancient live oak hammocks that form extensive forest as well as remnant islands of vegetation in open pastureland. These woodlands and pockets of deeper, rich soils, ample shade, and often springheads of moisture provide ideal habitats for several of the southeastern species of orchids. In addition to the broad, spreading live oaks, assorted species of magnolias and hollies are usually found. Depending on the amount of regular moisture, the hammocks may be hydric, mesic, or xeric.

Species typically found in the great temperate hammocks of the Southeast:

Corallorhiza odontorhiza var. *odontorhiza,* autumn coralroot
Corallorhiza wisteriana, Wister's coralroot
Epidendrum magnoliae, green-fly orchis
Gymnadeniopsis clavellata, little club-spur orchis
Habenaria odontopetala, toothed rein orchis
Habenaria quinqueseta, Michaux's orchid
Hexalectris spicata, crested coralroot
Listera australis, southern twayblade
Malaxis spicata, Florida adder's-mouth
Malaxis unifolia, green adder's-mouth
Platanthera flava var. *flava,* southern tubercled orchis

Platythelys querceticola, low ground orchid, jug orchid
Spiranthes odorata, fragrant ladies'-tresses
Spiranthes ovalis var. *ovalis,* southern oval ladies'-tresses
Spiranthes ovalis var. *erostellata,* northern oval ladies'-tresses
Spiranthes sylvatica, woodland ladies'-tresses
Tipularia discolor, crane-fly orchis
Triphora rickettii, Rickett's noddingcaps
Triphora trianthophora, three birds orchid

As in many habitats in the Southeast spring often begins in late winter with emerging plants of *Corallorhiza wisteriana,* Wister's coralroot, and *Listera australis,* the southern twayblade. The coralroot seems at home in either dry or mesic conditions but the twayblade requires a hydric environment. Both species are ephemeral; a month after flowering no sign of the plants is present until next season. The overhanging branches of the massive live oaks often support large colonies of *Epidendrum magnoliae,* the green-fly orchis, which may flower at any time of year. Occasionally the green-fly orchis will be found growing on maples, beeches, and junipers. *Spiranthes sylvatica,* the recently described woodland ladies'-tresses, is most at home in many of these hammocks and may appear in shaded, dry areas with little ground vegetation, in brushy thickets, or poking up along the border of the hammock and open fields.

In the late spring along the floodplains and streamsides within these hammocks, one can often find the slender inflorescences of *Platanthera flava* var. *flava,* the southern tubercled orchis, with their curious squared lips and delicious fragrance. Drier areas, along the eastern portions of our range, within the woodlands may have plants of the delicate *Malaxis unifolia,* the green adder's-mouth. This species also grows in many other areas. If encountered in open, dry areas, typified by reindeer lichen and haircap mosses, the plants should be carefully examined for the possibility of *M. bayardii,* Bayard's adder's-mouth.

Early summer brings the leafless stalks of *Hexalectris spicata,* the crested coralroot. This sticklike plant has striking and colorful flowers when observed close at hand.

Midsummer does not have an abundance of orchids flowering, but plants of *Tipularia discolor,* the crane-fly orchis, can often be found, and in a few selected hammocks *Triphora trianthophora,* the three birds orchid, may put on quite a show. In Columbia County, Florida, the Florida endemic *T. rickettii,* Rickett's noddingcaps, should be actively sought, per a report by Luer (1972) that has never been vouchered. The tiny plants with their tight, upright, little yellow flowers are certainly easy to overlook.

Although *Habenaria quinqueseta,* Michaux's orchis, is very rare north of Florida, midsummer is the time to search for this orchid with its spidery green and white flowers. Considered extirpated in Texas and South Carolina, it is known from only one county in southwestern Georgia, and is nearly as rare in Alabama, Mississippi, and Louisiana. In contrast, Michaux's orchis can be locally common throughout northern Florida.

In autumn, although in no way florally resembling *Habenaria quinqueseta, H. odontopetala,* the toothed rein orchis, is reaching the northern limit of its range in the Jacksonville, Florida, area. A large, thriving population occurs in a Jacksonville city park along the St. John's River. This is the northernmost known population for this essentially subtropical species. It can flower well into November.

The last of the autumn-flowering orchids usually includes *Corallorhiza odontorhiza,* the autumn coralroot, *Spiranthes odorata,* fragrant ladies'-tresses, and *S. ovalis,* the oval ladies'-tresses. Both the autumn coralroot and oval ladies'-tresses are woodland species and found in widely scattered and local colonies. Numbers may vary from only a few plants to more than a dozen.

Both species occur in two varieties. *Spiranthes ovalis* var. *erostellata,* the northern oval ladies'-tresses, is the more widespread variety and occurs in most states in the Southeast, whereas var. *ovalis,* the southern oval ladies'-tresses, is very rare, and found primarily in northern Florida, with a few sites in the southern portions of the Gulf states. *Corallorhiza odontorhiza* var. *odontorhiza,* the autumn coralroot, is the widespread variety, easily overlooked, and not well documented. *Corallorhiza odontorhiza* var. *pringlei,* Pringle's autumn coralroot, with fully opened, chasmogamous flowers, is known in the Southeast from a single herbarium record from northeastern Georgia. Both varieties grow in similar habitats within the hardwood forest.

5. Southern Appalachians

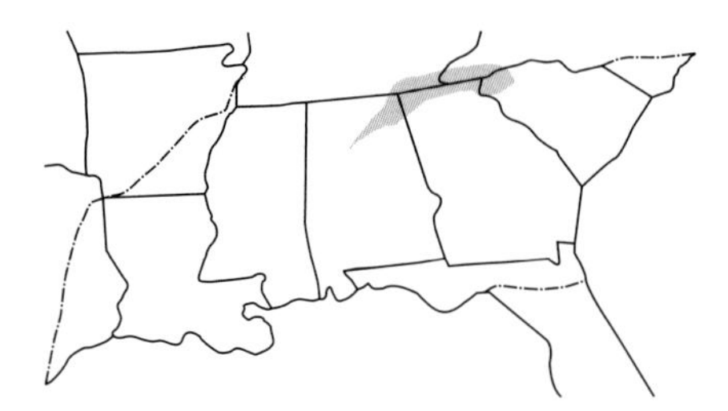

The massive Appalachian Mountain range, which begins in Newfoundland, Canada, reaches its southern terminus at Oak Mountain State Park just outside of Montgomery, Alabama. Within the Southeast, the Appalachians dip into northwestern South Carolina and northern Georgia, including Stone Mountain, the southern trailhead for the Appalachian Trail, and then make the gentle swing southwest toward Montgomery, Alabama. Lower, rolling hills are also evident in adjacent northern Mississippi. Many of the species of orchids found within these southern Appalachians are more typical of the central and northern states and reach their southern limit of distribution here. One of the most curious distribution patterns are enclaves of typically Coastal Plain species within the mountains. Flat Rock, near Hendersonville, North Carolina, is the most famous of these, but such areas also occur in northern South Carolina.

Species typically found in the great Southern Appalachians:

Aplectrum hyemale, putty-root
Arethusa bulbosa, dragon's-mouth
Cleistes bifaria, upland spreading pogonia
Corallorhiza maculata, spotted coralroot
Corallorhiza odontorhiza var. *odontorhiza,* autumn coralroot
Corallorhiza wisteriana, Wister's coralroot
Cypripedium acaule, pink lady's-slipper, moccasin flower

Cypripedium parviflorum var. *pubescens,* large yellow lady's-slipper
Epipactis helleborine, broad-leaved helleborine*
Galearis spectabilis, showy orchis
Goodyera pubescens, downy rattlesnake orchis
Gymnadeniopsis clavellata, little club-spur orchis
Isotria medeoloides, small whorled pogonia
Isotria verticillata, large whorled pogonia
Liparis liliifolia, lily-leaved twayblade
Liparis loeselii, Löesel's twayblade, fen orchis
Listera australis, southern twayblade
Listera smallii, Small's twayblade
Malaxis bayardii, Bayard's adder's-mouth
Malaxis unifolia, green adder's-mouth
Platanthera ciliaris, orange fringed orchis
Platanthera cristata, orange crested orchis
Platanthera flava var. *herbiola,* northern tubercled orchis
Platanthera grandiflora, large purple fringed orchis
Platanthera integrilabia, monkey-face orchis
Platanthera lacera, green fringed orchis, ragged orchis
[*Platanthera orbiculata,* pad-leaved orchis]
Platanthera peramoena, purple fringeless orchis
Platanthera psycodes, small purple fringed orchis
Spiranthes cernua, nodding ladies'-tresses
Spiranthes lacera var. *gracilis,* southern slender ladies'-tresses
Spiranthes magnicamporum, Great Plains ladies'-tresses
Tipularia discolor, crane-fly orchis
Triphora trianthophora, three birds orchid

Exploring this mountainous terrain is not always easy, but the cooler, lush forests certainly make up for the added exertion. Just following the roadsides always reveals some of the orchids, but taking the time to explore adjacent trails or simply wandering throughout the woodlands and glades is bound to reveal more species.

Spring within these southern mountains offers many species that continue their journey northward as the spring advances with time. *Corallorhiza wisteriana,* Wister's coralroot, and *Listera australis,* the southern twayblade, are usually the harbingers of spring and are soon followed by *Cypripedium acaule,* the pink lady's-slipper or moccasin flower, and *C. parviflorum* var. *pubescens,* the large yellow lady's-slipper. Not far away may be delightful stands of *Galearis spectabilis,* the showy orchis, and *Isotria verticillata,* the large whorled pogonia. Not documented for the Southeast, but occurring in bordering counties in North Carolina, *Coeloglossum viride,* the long-bracted green orchis, can be found. Often found nearby in similar habitats are the curious *Aplectrum hyemale,* the puttyroot, and *Liparis liliifolia,* the lily-leaved twayblade, both of which occupy rich woodlands. Although rare and local *Arethusa bulbosa,* the dragon's-mouth, *Cleistes bifaria,* the upland spreading pogonia, *Isotria medeoloides,* the federally threatened small whorled pogonia, and *Liparis loeselii,* Löesel's twayblade or fen orchis, also flower in late spring.

Early summer finds local scattered stands of *Corallorhiza maculata,* spotted coralroot, *Gymnadeniopsis clavellata,* the little club-spur orchis, *Listera smallii,* Small's twayblade, *Malaxis bayardii,* Bayard's adder's-mouth, *M. unifolia,* the green adder's-mouth, *Platanthera ciliaris,* the orange fringed orchis, *P. cristata,* the orange crested orchis, *P. grandiflora,* the large purple fringed orchis, *P. lacera,* the green fringed or ragged orchis, and *P. psycodes,* the small purple fringed orchis. The various species of fringed *Platanthera* are usually found on open mountain slopes and in glades, whereas the orange fringed and orange crested orchises are local disjuncts from the Coastal Plain.

Although *Platanthera flava* var. *herbiola,* the northern tubercled orchis, has been found in northern Georgia, it has also been reported recently from northwestern South Carolina, but verification remains to be made. *Platanthera orbiculata,* the pad-leaved orchis, has been reported in the literature from Georgia and occurs northward in the mountains of North Carolina. *Cypripedium reginae,* the showy lady's-slipper, is found just beyond the Georgia line in southern North Carolina. Although it is difficult to imagine this species being overlooked in either Georgia or South Carolina, the nearby location makes it a prime species to search for within the bordering counties.

Midsummer continues the array of orchids, and elevation often plays an important part in the flowering sequence. Depending on exactly where you are you might expect to find additional sites for *Epipactis helleborine,* broad-leaved helleborine*, a new arrival and currently known from a single site along an I-75 rest area, *Goodyera pubescens,* the downy rattlesnake orchis, a common component of the mountain forest floor, and *Platanthera integrilabia,* the monkey-face orchis, one of the rarest orchids to be found in North America. If you are adventuresome, look for it along the lower slopes of Tallulah Gorge, although it does occur in several other places in northern Georgia as well as in the Piedmont of Alabama and Mississippi. It may be extirpated from South Carolina. *Goodyera repens,* the lesser rattlesnake orchis, is another of the orchid trio that is found in the bordering counties of North Carolina and should be sought in northern Georgia and South Carolina as well.

Platanthera peramoena, the spectacular purple fringeless orchis, is another species no longer regularly found in South Carolina but it is well distributed in Georgia, Alabama, and especially northern and central Mississippi. Although *Spiranthes vernalis,* the grass-leaved ladies'-tresses, may be found in scattered areas at the lower elevations of the mountains in late spring, midsummer brings the widely distributed *S. lacera* var. *gracilis,* the southern slender ladies'-tresses, which favors mowed roadsides, old cemeteries, and even domestic lawns! Searching the deciduous woodlands for *Tipularia discolor,* the crane-fly orchis, at this time requires a carefully tuned eye; even though the plants are not small, the slender stalks with many dancing brown flowers tend to blend in with the surrounding vegetation. The most elusive species to see in flower is *Triphora trianthophora,* the three birds orchid. Never common, except in central Florida, this exquisite species is found in a full range of wooded habitats, but in the mountains it often favors American beech and southern hemlock woodlands. The three birds orchid flowers for only a few hours each day, blooming for only a few days, so catching it in bloom is a real challenge!

Autumn continues the parade of native orchids with three contrasting species. *Corallorhiza odontorhiza* var. *odontorhiza,* the autumn coralroot, is only a few centimeters tall and often flowers within the fallen leaves. It is a leafless species, which does not add to the ease of seeing it.

Two species of ladies'-tresses finish up the season. *Spiranthes cernua,* the nodding ladies'-tresses, often present in large numbers, is the most widespread of the genus in eastern North America. It can occur in a variety of habitats but usually favors damp, mown roadsides, old gravel pits, streambanks, etc.—just about anywhere that is grassy and wet. In contrast, *S. magnicamporum,* the Great Plains ladies'-tresses, is a prairie species found sparingly within glades and prairie-like areas in northeastern Georgia and south of the mountains in the Black Belt Prairie Regions of Alabama and Mississippi.

6. Bayou Country and the Mississippi Delta

The mighty Mississippi River is what feeds and renews this lush, rich habitat. Although much of the floodplain of the river has given way to cultivation, particularly rice, remnant woodlands and prairies still can be found in southern Arkansas and northern Louisiana. Further south in the Bayou Country, the intricate waterways and vast swamps harbor islands of land that still have a surprising number of orchids surviving. The habitat is not always inviting or hospitable. But, insects and reptiles aside, it does provide an abundance of acreage in which to search for wild orchids.

Species typically found in the Bayou Country and the Mississippi River Delta:

Calopogon barbatus, bearded grass-pink
Calopogon pallidus, pale grass-pink
Calopogon tuberosus, common grass-pink
Corallorhiza wisteriana, Wister's coralroot
Epidendrum magnoliae, green-fly orchis
Gymnadeniopsis integra, yellow fringeless orchis
Gymnadeniopsis nivea, snowy orchis
Habenaria quinqueseta, Michaux's orchid
Habenaria repens, water spider orchid
Listera australis, southern twayblade
Malaxis unifolia, green adder's-mouth

Platanthera ciliaris, orange fringed orchis
Platanthera conspicua, southern white fringed orchis
Platanthera cristata, orange crested orchis
Platanthera flava var. *flava,* southern tubercled orchis
Platythelys querceticola, low ground orchid, jug orchid
Pogonia ophioglossoides, rose pogonia
Ponthieva racemosa, shadow-witch
Pteroglossaspis ecristata, crestless plume orchid
Spiranthes laciniata, lace-lipped ladies'-tresses
Spiranthes longilabris, long-lipped ladies'-tresses
Spiranthes odorata, fragrant ladies'-tresses
Spiranthes praecox, giant ladies'-tresses
Spiranthes sylvatica, woodland ladies'-tresses
Spiranthes tuberosa, little ladies'-tresses
Spiranthes vernalis, grass-leaved ladies'-tresses

In addition to covering this area seasonally, we will try to look at the orchids from the degree of rarity. Fifteen species are frequently seen in the Mississippi basin. Nine of these—*Calopogon tuberosus,* the common grass-pink, *Corallorhiza wisteriana,* Wister's coralroot, *Listera australis,* the southern twayblade, *Pogonia ophioglossoides,* rose pogonia, *Ponthieva racemosa,* the shadow-witch, *Spiranthes odorata,* the fragrant ladies'-tresses, *S. praecox,* the giant ladies'-tresses, *S. sylvatica,* the woodland ladies'-tresses, and *S. vernalis,* the grass-leaved ladies'-tresses—are all seen in such abundance that they can be termed common in appropriate habitat. The coralroot, twayblade, and shadow-witch are woodland species and often accompanied by the woodland and fragrant ladies'-tresses.

Late winter and spring are the time for Wister's coralroot and southern twayblade, soon to be followed by the spring trio of ladies'-tresses: *Spiranthes praecox, S. sylvatica,* and *S. vernalis.* Later in the spring in the open meadows and shorelines the common grass-pink and rose pogonia brighten up the landscape. The rich soils of the wooded floodplains often are a suitable home for *Platanthera flava* var. *flava,* the southern tubercled orchis, and by autumn *Ponthieva* and *S. odorata,* the fragrant ladies'-tresses, may be found in many of the same areas.

Epidendrum magnoliae, the green-fly orchis, is the only epiphyte to be found in Louisiana and favors trunks and branches of old live oaks, beeches, and southern magnolia. It flowers sporadically throughout the year.

Early summer brings *Gymnadeniopsis nivea,* the snowy orchis, and then an array of common orchids continues with *Habenaria repens,* the water spider orchid, our only aquatic orchid, and both *Platanthera ciliaris,* the orange fringed orchis, and *P. cristata,* the orange crested orchis. The two orange cousins are found in a variety of habitats, from the edges of swampy ground to roadside ditches. Although growing in much drier areas, *Spiranthes tuberosa,* the little ladies'-tresses, is well distributed throughout the area.

Some of the rarities to search for in the spring might include *Calopogon barbatus,* the bearded grass-pink, and the similar *C. oklahomensis,* the Oklahoma grass-pink, both found in remnant prairie areas. *Calopogon pallidus,* the pale grass-pink, is scattered around, often in the same areas as its commoner cousin *C. tuberosus.*

Midsummer brings on five of the rarest orchids in Louisiana. Although none are really bayou species, these can be found within the floodplain and historic prairies along the Mississippi River: *Gymnadeniopsis integra,* the yellow fringeless orchis, *Habenaria quinqueseta,* Michaux's orchid, *Platanthera conspicua,* the southern white fringed orchis, *Platythelys querceticola,* the low ground orchid, and *Pteroglossaspis ecristata,* the crestless plume orchid. All are known from only a few locales and usually only a few plants. All prefer open, damp areas, with the exception of the *Platythelys,* whose habitat is shaded river runs.

7. Lowcountry

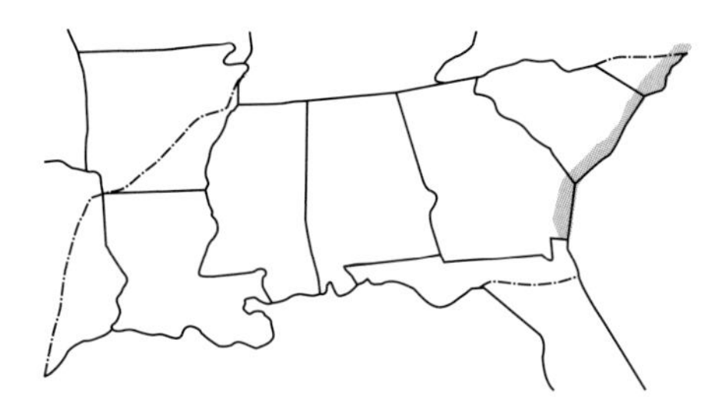

The eastern Atlantic outer Coastal Plain, usually from central Georgia to southern North Carolina, is often referred to as the Lowcountry. Much of the land is at or near (or below) sea level and the winter weather is very moderate. Extensive coastal hammocks abound and although there has been too much development, there are still large tracts of land in private and public holdings. Many rivers such as the Savannah and Peedee find their way to the ocean through this area. Salt marshes and estuaries marble the coastline, and on some of the higher ground ancient live oak hammocks still thrive. The Francis Marion National Forest just north of Charleston is one of the largest public lands in this area and offers almost unlimited orchid hunting. So much so that one of the roads is locally dubbed "Orchid Alley"! The Lowcountry region as well as the Southeast as defined in this work ends at the Wilmington, North Carolina, area with Lake Waccamaw and the Green Swamp, both significant orchid areas.

One of the characteristic landforms throughout the Lowcountry is the Carolina Bay. These "lakes" were formed many centuries ago and are shallow bodies of water that vary from small ponds to the massive Lake Waccamaw. In many of the bays the water level can fluctuate annually; in some years it is deep and in others shallow or nearly dry. Although not necessarily always the best of orchid habitats, these Carolina Bays do account for habitat for the majority of rare, threatened, and endangered species found along the Coastal Plain. Other critical landforms, especially for orchids, are the poccasins and savannas that are found throughout the Coastal Plain and Lowcountry.

Species typically found in the Lowcountry and Atlantic Coastal Plain:

Calopogon barbatus, bearded grass-pink
Calopogon multiflorus, many-flowered grass-pink
Calopogon pallidus, pale grass-pink
Calopogon tuberosus, common grass-pink
Cleistes bifaria, upland spreading pogonia
Cleistes divaricata, spreading pogonia
Corallorhiza wisteriana, Wister's coralroot
Epidendrum magnoliae, green-fly orchis
Gymnadeniopsis integra, yellow fringeless orchis
Gymnadeniopsis nivea, snowy orchis
Habenaria quinqueseta, Michaux's orchid
Habenaria repens, water spider orchid
Malaxis spicata, Florida adder's-mouth
Platanthera ciliaris, orange fringed orchis
Platanthera conspicua, southern white fringed orchis
Platanthera cristata, orange crested orchis
Platanthera flava var. *flava,* southern tubercled orchis
Pogonia ophioglossoides, rose pogonia
Ponthieva racemosa, shadow-witch
Pteroglossaspis ecristata, crestless plume orchid
Spiranthes cernua, nodding ladies'-tresses
Spiranthes eatonii, Eaton's ladies'-tresses
Spiranthes floridana, Florida ladies'-tresses
Spiranthes laciniata, lace-lipped ladies'-tresses
Spiranthes longilabris, long-lipped ladies'-tresses
Spiranthes odorata, fragrant ladies'-tresses
Spiranthes ovalis var. *erostellata,* northern oval ladies'-tresses
Spiranthes praecox, giant ladies'-tresses
Spiranthes sylvatica, woodland ladies'-tresses
Spiranthes tuberosa, little ladies'-tresses
Spiranthes vernalis, grass-leaved ladies'-tresses

Although not as abundant as they may be further inland from the coast, plants of *Corallorhiza wisteriana,* Wister's coralroot, signal the beginning of the orchid year. Usually found in dry hammocks, they can be found in many of the local county and city parks. *Spiranthes eatonii,* Eaton's ladies'-tresses and *S. floridana,* the Florida ladies'-tresses, are two of the earliest flowering of the ladies'-tresses and, based upon recent collections, the former is apparently widespread in the Charleston area. Old, thin-soil cemeteries can often be an excellent habitat for this recently described species. *Spiranthes floridana,* one of the rarest orchids in North America, has several records for the Lowcountry, but no recent reports. It should be carefully sought.

The most common ladies'-tresses, *Spiranthes vernalis,* the grass-leaved ladies'-tresses, a favorite of lawns and roadsides, soon follows and remains in flower for nearly two months. In the damper areas, pine flatwoods, and savannas *S. praecox,* the giant ladies'-tresses, with its pure white tubular flowers and green-veined lips, will soon be waving its slender spikes. Recent reports of *S. sylvatica,* the woodland ladies'-tresses, have come from the Savannah River area and the Green Swamp to the north. Excellent herbarium records from The Citadel in Charleston indicate that it is locally common not far from the city.

Pond shores, damp pine flatwoods, and roadside seeps throughout the area can sport populations of *Calopogon barbatus,* the bearded grass-pink. Nowhere common, it is best seen at the Green Swamp, where the white-flowered form is also present. Spring is also the season to search recently burned flatwoods for the globally rare *C. multiflorus,* the many-flowered grass-pink. A recently discovered population has been seen not far from Charleston. Soon to follow, and more frequent, will be *C. pallidus,* the pale grass-pink, and *C. tuberosus,* the common grass-pink. Two of the choicest of our spring-flowering orchids, *Cleistes bifaria,* the upland spreading pogonia, and the larger flowered *C. divaricata,* the spreading pogonia, flower in mid- to late May. Both species may be been seen near the Big Savannah, Brunswick, Georgia, and at the Green Swamp in North Carolina. In other places it is usually one species or the other, with *C. divaricata* being the more common of the two on the Coastal Plain.

Occurring almost anywhere in a moist habitat is the delightful *Pogonia ophioglossoides,* the rose pogonia. These charming elves of the orchid world are fairly frequent in a wide variety of both geographic and wetland habitats.

The advent of summer slows down the number of orchid species seen, but more than makes up for it in many large, showy species. *Gymnadeniopsis nivea,* the snowy orchis, is a frequently seen species along US 301 in southern Georgia in early June, where earlier *Calopogon barbatus* was locally abundant. Scattered within the snowy orchids may be plants of *Spiranthes laciniata,* the lace-lipped ladies'-tresses. The latter species, as well as the snowy orchis, will be found northward all the way to the Green Swamp area. *Platanthera flava* var. *flava,* the southern tubercled orchis, is occasionally found along roadside ditches, but is more at home in the wet, wooded swamps. Although not the showiest of species, it is one of the most fragrant.

The showiest of the summer-flowering orchids are a trio consisting of *Pla-*

tanthera ciliaris, the orange fringed orchis, *P. conspicua,* the southern white fringed orchis, and *P. cristata,* the orange crested orchis. The three can be found in Francis Marion National Forest along some of the wildflower trails, as well as in many other places throughout the Coastal Plain and Lowcountry. Both orange-flowered species range well inland beyond the Coastal Plain but *P. conspicua,* the southern white fringed orchis, is restricted to a narrow strip from southeastern North Carolina southwestward to southern Georgia. Further inland, especially in the Piedmont, *P. blephariglottis,* the northern white fringed orchis, reaches the southern limit of its range. If you come across plants of white fringed orchises examine them carefully to determine which species you have found.

Searching overhead in the large, moss- and fern-laden live oaks should reveal flowering plants of *Epidendrum magnoliae,* the green-fly orchis, which reaches the northern limit of its range in the Wilmington, North Carolina, area. It is the only epiphytic orchid to be found north of Florida.

The drier roadsides and, again, often cemeteries may host plants of *Spiranthes tuberosa,* the little ladies'-tresses. These pristine white-flowered gems have no leaves present at flowering time, and may be locally abundant throughout the entire Coastal Plain.

Two of the rarest species found in the Lowcountry are flowering in August. *Gymnadeniopsis integra,* the yellow fringeless orchis, a species that is rapidly disappearing throughout its range, is known from Francis Marion National Forest in open savannas and damp roadsides. Although the plants flower infrequently, they are often victim to roadside mowing practices.

Habenaria quinqueseta, Michaux's orchid, reaches the northern limit of its range in five counties in southeastern South Carolina but has not been seen in many years and is currently considered extirpated. In Georgia it is known from a single site, but in Florida is considerably more frequent. For much of August and September the only aquatic orchid we have, *Habenaria repens,* the water spider orchid, will be flowering in roadside ditches, diked waterways, and shallow Carolina Bays. It can often form floating colonies of several hundred plants. This is also the time to watch the pine flatwoods and hedgerows for plants of the curious *Pteroglossaspis ecristata,* the crestless plume orchid, with its green and black (or purple) orchids on a stick. This species is considered local and rare, and colonies vary greatly in numbers and dependability of flowering from year to year.

Late summer and early autumn in the wooded swamps are the ideal times to search for *Malaxis spicata,* the Florida adder's-mouth, and *Ponthieva racemosa,* the shadow-witch. They often occupy the same habitat niches, although the shadow-witch will grow in drier areas. Many of these same areas will have *Spiranthes odorata,* fragrant ladies'-tresses, later in the season.

By late September and early October a few rich woodlands may have plants of the rare *Spiranthes ovalis* var. *erostellata,* the northern oval ladies'-tresses. This is a species that seems to be increasing, especially in recent second-growth hardwoods. The tiny, pure white flowers are held in three ranks that have a decided corkscrew appearance when viewed from above. *Spiranthes longilabris,* the long-lipped ladies'-tresses, is usually the latest to flower of the ladies'-tresses, and of the native orchids in this area. Thought by many to be the most handsome of the genus in eastern North America, the stately spikes present their large flowers in a single rank that shows off each individual bloom to its best advantage. This is a species primarily of open pine flatwoods and savannas. It is becoming increasingly rare throughout most of its range, although the Green Swamp area still boasts several good colonies.

8. Piedmont Hills

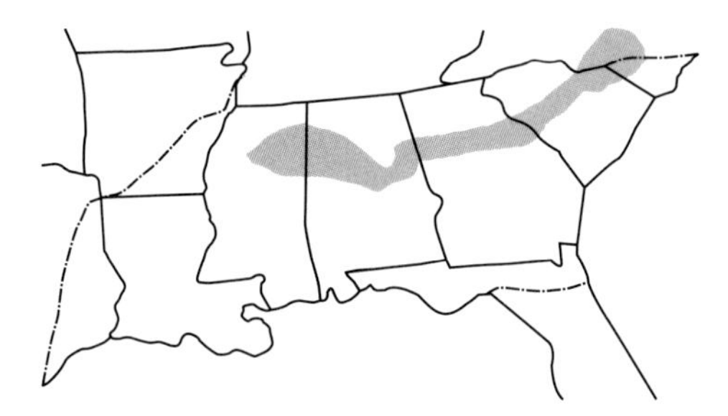

The Piedmont is that wonderful upland area that bridges the gap between the Coastal Plain and the mountains. It can be flat and sandy in areas or steep and wooded. Sandhills abound, especially in South Carolina, and the lush foothills of the Appalachians are home to some of the northern species. Because this is a transitional area, the species list is longer than some areas. Coastal Plain islands occur as well as higher hills, and a fine line makes dividing the zones difficult in many places. Soils tend to be less sandy and richer in the hills and support a tree and shrub layer that is not usually found on the Coastal Plain. Because of the higher elevations, winter, such as it is, comes in interesting pockets of ice and snow some years and not a flake in others. Plants in the Piedmont need to be just a bit tougher to accommodate the coolness of winter and the heat of summer.

Species typically found within the Piedmont:

Calopogon oklahomensis, Oklahoma grass-pink
Calopogon tuberosus, common grass-pink
Corallorhiza odontorhiza var. *odontorhiza,* autumn coralroot
Corallorhiza odontorhiza var. *pringlei,* Pringle's autumn coralroot
Corallorhiza wisteriana, Wister's coralroot
Cypripedium kentuckiense, ivory-lipped lady's-slipper
Cypripedium parviflorum var. *parviflorum,* southern small yellow lady's-slipper

Cypripedium parviflorum var. *pubescens,* large yellow lady's-slipper
Galearis spectabilis, showy orchis
Gymnadeniopsis clavellata, little club-spur orchis
Hexalectris spicata, crested coralroot
Isotria medeoloides, small whorled pogonia
Isotria verticillata, large whorled pogonia
Liparis liliifolia, lily-leaved twayblade
Liparis loeselii, Löesel's twayblade
Listera australis, southern twayblade
Malaxis unifolia, green adder's-mouth
Platanthera blephariglottis, northern white fringed orchis
Platanthera ciliaris, orange fringed orchis
Platanthera cristata, orange crested orchis
Spiranthes cernua, nodding ladies'-tresses
Spiranthes magnicamporum, Great Plains ladies'-tresses
Spiranthes ovalis var. *ovalis,* southern oval ladies'-tresses
Spiranthes ovalis var. *erostellata,* northern oval ladies'-tresses
Spiranthes vernalis, grass-leaved ladies'-tresses
Tipularia discolor, crane-fly orchis
Triphora trianthophora, three birds orchid

Spring begins in the Piedmont much as it does throughout the Southeast with the appearance of *Corallorhiza wisteriana,* Wister's coralroot, and *Listera australis,* the southern twayblade. Both are plants of mixed woodlands and the *Listera* especially likes springheads. In a few remnant prairie habitats, plants of *Calopogon oklahomensis,* the Oklahoma grass-pink, should be sought. The South Carolina record is not specific, but Doug Goldman (pers. comm.) has hypothesized that it most likely occured within the sandhills. The open deciduous woodlands may present flowering plants of *Isotria verticillata,* the large whorled pogonia, and *Cypripedium acaule,* the pink lady's-slipper, especially in the Dothan, Alabama, area. In a very few places in South Carolina and Georgia the federally threatened *Isotria medeoloides,* the small whorled pogonia, has recently been found.

As spring progresses plants of *Malaxis unifolia,* the green adder's-mouth, *Liparis liliifolia,* the lily-leaved twayblade, and, in calcareous areas, *Galearis spectabilis,* the showy orchis, will begin flowering. Although known only from a few sites in Mississippi and Georgia, *L. loeselii,* Löesel's twayblade, is worth searching for at this southern limit of its range. Damp grassy roadsides, old borrow pits, and open woodlands are the preferred habitat. That old perennial warhorse, *Spiranthes vernalis,* the grass-leaved ladies'-tresses, seems equally at home in the Piedmont as on the Coastal Plain. It is vouchered from nearly every county in the Southeast!

Although never really abundant in the Southeast except for East Texas, Louisiana, and adjacent Arkansas, *Cypripedium kentuckiense,* the ivory-lipped lady's-slipper, is a spectacular sight on damp slopes and floodplains in rich woods. It is found in the Southeast in small numbers in all the other states except South Carolina and Florida. The familiar yellow lady's-slipper is present in two varieties in the Southeast. *Cypripedium parviflorum* var. *parviflorum,* the southern small yellow lady's-slipper, a plant of more acidic woodlands, is rare and local but may be more widespread than records indicate. *Cypripedium parviflorum* var. *pubescens,* the large yellow lady's-slipper, is the most frequently encountered yellow-flowered variety in the Southeast. It prefers rich, limy woodlands and can also be found in sunny, calcareous areas. It is present in the Piedmont and mountains of much of the Southeast.

Early summer is the time to search for *Gymnadeniopsis clavellata,* the little club-spur orchis. This green-flowered rein orchis is often found in the central and northern counties of the Southeast along stream runs. In the drier woodlands *Hexalectris spicata,* the crested coralroot, springs forth with its beautiful striped and crested flowers on leafless stems, some to nearly a meter in height!

By midsummer the time has come to watch for the trio of *Platanthera blephariglottis,* the northern white fringed orchis, *P. ciliaris,* the orange fringed orchis, and *P. cristata,* the orange crested orchis. It is not at all unusual to find hybrids among these three and the flowers are often paler yellow or creamy coffee-colored. Forsaking the roadsides for open, deciduous woodlands may enable the explorer to find *Tipularia discolor,* the crane-fly orchis, and, for the very fortunate, the delightful and elusive *Triphora trianthophora,* the three birds orchid. Remember that this diminutive gem of the forest flowers for only a few hours and a few days of the year!

Fall is fast approaching and a selection of ladies-tresses ushers in the shorter, cooler days. *Spiranthes cernua,* the nodding ladies'-tresses, is the most widespread and often grows in large numbers along damp roadsides. The typical prairie species *S. magnicamporum,* the Great Plains ladies'-tresses, is surprisingly well distributed in northern Georgia, central Alabama, and Mississippi, as well as west to Louisiana and the remnant prairies of eastern Arkansas.

Leaving the open grassland for the cool woodlands may lead the searcher to occasional plants of *Spiranthes ovalis* var. *ovalis,* the southern oval ladies'-tresses, but more often to *S. ovalis* var. *erostellata,* the northern oval ladies'-tresses. Both varieties are known from most of the southeastern states except South and North Carolina, where only the latter variety is present. Woodlands similar to the oval ladies'-tresses' habitat may also hide the small plants of *Corallorhiza odontorhiza* var. *odontorhiza,* the autumn coralroot. The plants often flower among the falling leaves, and their leafless stems blend in with the surrounding vegetation, making them difficult to find. Known in the Southeast from only a single record in northeastern Georgia, *C. odontorhiza* var. *pringlei,* Pringle's autumn coralroot, is a variety with fully opened, showy flowers.

9. At the Limit

A good many species of native orchids reach their distributional limits in the southeastern United States. The majority of these are northern species reaching their southern limits, and a few are extralimital with decidedly disjunct populations. Several of these more northerly species follow the Appalachian Mountains southward while a few subtropical species found primarily from central Florida south reach the northern limit of their continuous range.

Eulophia alta, the wild coco orchid, *Mesadenus lucayanus,* the copper ladies'-tresses, *Sacoila lanceolata,* the leafless beaked orchid, *Triphora rickettii,* Rickett's noddingcaps, and *Habenaria odontopetala,* the toothed rein orchis, are all examples of Floridian or subtropical species found well distributed throughout central and/or southern Florida (and many to the islands and well beyond), reaching the northern extreme of their distribution in the northern counties of Florida or nearby Georgia. *Habenaria quinqueseta,* Michaux's orchid, also reaches its northern limit but is distributed, at least historically, north to South Carolina and west to Texas.

Five species found primarily in the central Atlantic and Midwestern states find their way into the Southeast. *Platanthera peramoena,* the purple fringeless orchis, and *Spiranthes magnicamporum,* the Great Plains ladies'-tresses, both have a contiguous and natural extension southeastward, although the latter has recently been found in southwestern Virginia, which may represent a disjunct population. *Calopogon oklahomensis,* the Oklahoma grass-pink, presents an interesting situation regarding its distribution. Originally thought to be centered in the southern prairie states, it is now known, albeit from herbarium records, to have once ranged into Alabama, Mississippi, Georgia, and South Carolina. The two small colonies of *Cypripedium candidum,* the small white lady's-slipper, in central and northern Alabama represent disjunct populations of this north-central prairie species. The single recent record for *Spiranthes lucida,* the shining ladies'-tresses, also from central Alabama is another disjunct population. In central

Mississippi a colony of *Liparis loeselii,* Löesel's twayblade or fen orchis, is two states away from its nearest site to the north.

The historical record of *Platanthera leucophaea,* the eastern prairie fringed orchis, in Louisiana is also a disjunct. The site not only represents the southern limit of its range but one of the very few records from west of the Mississippi River.

The many species that travel south in the Appalachians often terminate their ranges in scattered populations. Such species as *Goodyera pubescens,* the downy rattlesnake orchis, *Gymnadeniopsis clavellata,* the little club-spur orchis, *Isotria verticillata,* the large whorled pogonia, *Cypripedium acaule,* the pink lady's-slipper or moccasin flower, and *Platanthera lacera,* the green fringed or ragged orchis, all have scattered populations down through the central and southern counties of several of the southeastern states. The first three reach to the central Panhandle of Florida near the Georgia/Alabama border where the more northern affinity soils are present. Two additional rarities, *Malaxis bayardii,* Bayard's adder's-mouth (in South Carolina), and *Isotria medeoloides,* the small whorled pogonia (in Georgia and South Carolina), both species that are widely scattered along the Lower Great Lakes, Appalachian Mountains, and eastern seaboard states, have isolated populations at their limits in the Southeast.

Only a few truly northern species are found this far south in the Southeast, and they are almost always in the mountains. This would include *Arethusa bulbosa,* the dragon's-mouth, *Platanthera grandiflora,* the large purple fringed orchis, *P. psycodes,* the small purple fringed orchis, and *Spiranthes ochroleuca,* the yellow ladies'-tresses. Their populations are few and all become more abundant as one travels north in the mountains.

Platanthera blephariglottis, the northern white fringed orchis, curiously enough reaches its southern limit within the Piedmont rather than in the mountains of Georgia, whereas *Listera smallii,* Small's twayblade, is a strictly Appalachian species found from New Jersey and Pennsylvania south in the mountains, terminating its distribution in northern Georgia.

10. Tips and Trips

Orchid hunting—but not picking!—in the southeastern United States can be both exciting and rewarding. Whether hunting for a few hours, a few days, or a few weeks, one is bound to find plentiful and exciting orchids just about everywhere! If you have only a few hours to spare, many rural roadsides in the Southeast will yield grass-pinks, ladies'-tresses, and fringed orchises. Time of year is critical, of course, but most anytime except for the dead of winter (and sometimes that can be productive as well), will guarantee you at least four or five species in flower within an hour's drive. If you are in the Jacksonville, Florida, area be sure to visit one of the many city parks. Remember that Jacksonville encompasses the entirety of Duval County so the county and city park system are the same. Also Ft. George State Park east of Jacksonville is always worth a stop as it is the northernmost known site for the copper ladies'-tresses, *Mesadenus lucayanus,* which should be in flower in February. For state forests, Cary State Forest on US 301 in May can often produce the exquisite large rosebud orchid, *Cleistes divaricata.* The ancient live oaks harbor many plants of the green-fly orchid, *Epidendrum magnoliae,* and that can flower at most any time of year. Other cities in the Southeast such as Charleston, Savannah, Columbia, Valdosta, and of course New Orleans all have choice habitat so be sure to check out those parks and state forests when you visit.

For those with a bit more time, an exciting week's trip goes from northern Florida to eastern Arkansas at the end of April or early May. Again, by stopping at the local state parks and watching the roadsides carefully you should be able to see four ladies'-tresses: *Spiranthes eatonii,* Eaton's ladies'-tresses, *S. vernalis,* the grass-leaved ladies'-tresses, *S. praecox,* the giant ladies'-tresses, and *S. sylvatica,* the woodland ladies'-tresses; three grass-pinks: *Calopogon tuberosus,* the common grass-pink, *C. pallidus,* the pale grass-pink, and if you are very fortunate, *C. multiflorus,* the many-flowered grass-pink, as well as *Pogonia ophioglossoides,* the rose pogonia.

After you cross the Mississippi River in northeastern Mississippi you will shortly arrive at the remnant prairies of eastern Arkansas and the goal of your quest, the Oklahoma grass-pink, *Calopogon oklahomensis.* Growing within the prairie and accompanied by a tapestry of spring prairie wildflowers, this recently described species should be in bloom. Among them will be the scarlet painted cups, *Castelleja* sp., golden coreopsis, *Coreopsis* spp., various phlox in shades of magenta and purple, green prairie milkweeds, and an abundance of grasses. These narrow remnants are surrounded by rice fields and are what little is left of the original terrain. Take the secondary roads through Alabama and Mississippi for a real treat as you pass through the Piedmont and into the southern extreme of the Appalachian Mountains in Montgomery. Be sure to visit Oak Mountain State Park on the north side of the city, as this is the actual terminus of the mountain range. Yellow lady's-slippers and showy orchises as well as the putty-root and downy rattlesnake orchid all make their home there.

Appendix 1. Distribution of the Wild Orchids of the Southeastern United States

Distribution of the Wild Orchids of the Southeastern United States

Aplectrum hyemale, putty-root, Adam and Eve
Arethusa bulbosa, dragon's-mouth
Bletilla striata, urn orchid*
Calopogon barbatus, bearded grass-pink
Calopogon multiflorus, many-flowered grass-pink
Calopogon oklahomensis, Oklahoma grass-pink
Calopogon pallidus, pale grass-pink
Calopogon tuberosus, common grass-pink
Cleistes bifaria, upland spreading pogonia
Cleistes divaricata, spreading pogonia
Corallorhiza maculata, spotted coralroot
Corallorhiza odontorhiza var. *odontorhiza*, autumn coralroot
Corallorhiza odontorhiza var. *pringlei*, Pringle's autumn coralroot
Corallorhiza wisteriana, Wister's coralroot
Cypripedium acaule, pink lady's-slipper, moccasin flower
Cypripedium candidum, small white lady's-slipper
Cypripedium kentuckiense, ivory-lipped lady's-slipper
Cypripedium parviflorum var. *parviflorum*, southern small yellow lady's-slipper
Cypripedium parviflorum var. *pubescens*, large yellow lady's-slipper
Epidendrum magnoliae, green-fly orchis
Epipactis helleborine, broad-leaved helleborine*
Eulophia alta, wild coco
Galearis spectabilis, showy orchis
Goodyera pubescens, downy rattlesnake orchis
Gymnadeniopsis clavellata, little club-spur orchis
Gymnadeniopsis integra, yellow fringeless orchis
Gymnadeniopsis nivea, snowy orchis
Habenaria odontopetala, toothed habenaria
Habenaria quinqueseta, Michaux's orchid
Habenaria repens, water spider orchid
Hexalectris spicata, crested coralroot
Isotria medeoloides, small whorled pogonia
Isotria verticillata, large whorled pogonia
Liparis liliifolia, lily-leaved twayblade
Liparis loeselii, Loesel's twayblade, fen orchis
Listera australis, southern twayblade
Listera smallii, Small's twayblade

AL	AR	GA	FL	LA	MS	NC	SC	TX
X		X					X	
							X	
			X					
X		X	X	X	X	X	X	
X		X	X	X	X	X	X	
X	X	X		X	X	X	X	X
X		X	X	X	X	X	X	
X	X	X	X	X	X		X	X
X		X	X		X	X	X	X
		X	X			X	X	
		X					X	
X	X	X	X	X	X	X	X	X
		X						
X	X	X	X	X	X	X	X	X
X		X			X	X	X	
X								
X	X	X		X	X			X
X		X						
		X			X		X	
X		X	X	X	X	X	X	
		X						
		X	X					
X	X	X			X		X	
X	X	X	X		X	X	X	
X	X	X	X	X	X	X	X	X
X		X	X	X	X	X	X	X
X	X	X	X	X	X	X	X	X
			X					
X		X	X	X	X	X	X	X
X	X	X	X	X	X	X	X	X
X	X	X	X	X	X	X	X	X
		X					X	
X	X	X	X	X	X	X	X	X
X		X			X		X	
X					X			
X	X	X	X	X	X	X	X	X
		X					X	

continued

Malaxis bayardii, Bayard's adder's-mouth
Malaxis spicata, Florida adder's-mouth
Malaxis unifolia, green adder's-mouth
Mesadenus lucayanus, copper ladies'-tresses
Platanthera blephariglottis, northern white fringed orchis
Platanthera chapmanii, Chapman's fringed orchis
Platanthera ciliaris, orange fringed orchis
Platanthera conspicua, southern white fringed orchis
Platanthera cristata, orange crested orchis
Platanthera flava var. *flava*, southern tubercled orchis
Platanthera flava var. *herbiola*, northern tubercled orchis
Platanthera grandiflora, large purple fringed orchis
Platanthera integrilabia, monkey-face orchis
Platanthera lacera, green fringed orchis, ragged orchis
Platanthera leucophaea, eastern prairie fringed orchis
Platanthera peramoena, purple fringeless orchis
Platanthera psycodes, small purple fringed orchis
Platythelys querceticola, low ground orchid, jug orchid
Pogonia ophioglossoides, rose pogonia
Ponthieva racemosa, shadow-witch
Pteroglossaspis ecristata, crestless plume orchid
Sacoila lanceolata, leafless beaked orchid
Spiranthes brevilabris, short-lipped ladies'-tresses
Spiranthes cernua, nodding ladies'-tresses
Spiranthes eatonii, Eaton's ladies'-tresses
Spiranthes floridana, Florida ladies'-tresses
Spiranthes lacera var. *gracilis*, southern slender ladies'-tresses
Spiranthes laciniata, lace-lipped ladies'-tresses
Spiranthes longilabris, long-lipped ladies'-tresses
Spiranthes lucida, shining ladies'-tresses
Spiranthes magnicamporum, Great Plains ladies'-tresses
Spiranthes ochroleuca, yellow ladies'-tresses
Spiranthes odorata, fragrant ladies'-tresses
Spiranthes ovalis var. *ovalis*, southern oval ladies'-tresses
Spiranthes ovalis var. *erostellata*, northern oval ladies'-tresses
Spiranthes parksii, Navasota ladies'-tresses
Spiranthes praecox, giant ladies'-tresses
Spiranthes sylvatica, woodland ladies'-tresses

AL	AR	GA	FL	LA	MS	NC	SC	TX
							X	
		X	X			X	X	
X	X	X	X	X	X	X	X	X
			X					
		X				X	X	
		X	X					X
X	X	X	X	X	X	X	X	X
X		X	X	X	X	X	X	X
X	X	X	X	X	X	X	X	X
X	X	X	X	X	X	X	X	X
		X						
		X						
X		X		X	X		X	
X	X	X		X	X	X	X	X
				x				
X		X			X		X	
		X		X			X	
			X	X	X			
X	X	X	X	X	X	X	X	X
X		X	X	X	X	X	X	X
X		X	X		X	X	X	
			X					
			X		X			X
X	X	X	X	X	X	X	X	X
X		X	X	X	X	X	X	X
X		X	X	X	X	X	X	X
X	X	X		X	X	X	X	X
X		X	X	X	X	X	X	X
X		X	X	X	X	X	X	X
X								
X	X	X		X	X			X
							X	
X	X	X	X	X	X	X	X	X
X	X	X	X	X	X			
X	X	X	X	X	X	X	X	X
								X
X	?	X	X	X	X	X	X	X
X	X	X	X	X	X	X	X	X

continued

Spiranthes tuberosa, little ladies'-tresses
Spiranthes vernalis, grass-leaved ladies'-tresses
Tipularia discolor, crane-fly orchis
Triphora rickettii, Rickett's noddingcaps
Triphora trianthophora, three birds orchid
Zeuxine strateumatica, lawn orchid

AL	AR	GA	FL	LA	MS	NC	SC	TX
X	X	X	X	X	X	X	X	X
X	X	X	X	X	X	X	X	X
X	X	X	X	X	X	X	X	X
			X					
X	X	X	X	X	X	X	X	X
X		X	X	X	X		X	X

Appendix 2. Flowering Times for the Wild Orchids of the Southeastern United States

Flowering Times for the Wild Orchids of the Southeastern United States

Aplectrum hyemale, putty-root
Arethusa bulbosa, dragon's-mouth
Bletilla striata, urn orchid*
Calopogon barbatus, bearded grass-pink
Calopogon multiflorus, many-flowered grass-pink
Calopogon oklahomensis, Oklahoma grass-pink
Calopogon pallidus, pale grass-pink
Calopogon tuberosus, common grass-pink
Cleistes bifaria, upland spreading pogonia
Cleistes divaricata, spreading pogonia
Corallorhiza maculata, spotted coralroot
Corallorhiza odontorhiza var. *odontorhiza*, autumn coralroot
Corallorhiza odontorhiza var. *pringlei*, Pringle's autumn coralroot
Corallorhiza wisteriana, Wister's coralroot
Cypripedium acaule, pink lady's-slipper, moccasin flower
Cypripedium candidum, small white lady's-slipper
Cypripedium kentuckiense, ivory-lipped lady's-slipper
Cypripedium parviflorum var. *parviflorum*, southern small yellow lady's-slipper
Cypripedium parviflorum var. *pubescens*, large yellow lady's-slipper
Epidendrum magnoliae, green-fly orchis
Epipactis helleborine, broad-leaved helleborine*
Eulophia alta, wild coco
Galearis spectabilis, showy orchis
Goodyera pubescens, downy rattlesnake orchis
Gymnadeniopsis clavellata, little club-spur orchis
Gymnadeniopsis integra, yellow fringeless orchis
Gymnadeniopsis nivea, snowy orchis
Habenaria odontopetala, toothed habenaria
Habenaria quinqueseta, Michaux's orchid
Habenaria repens, water spider orchid
Hexalectris spicata, crested coralroot
Isotria medeoloides, small whorled pogonia
Isotria verticillata, large whorled pogonia
Liparis liliifolia, lily-leaved twayblade
Liparis loeselii, Löesel's twayblade, fen orchis
Liparis x*jonesii*, Jones' hybrid twayblade
Listera australis, southern twayblade
Listera smallii, Small's twayblade

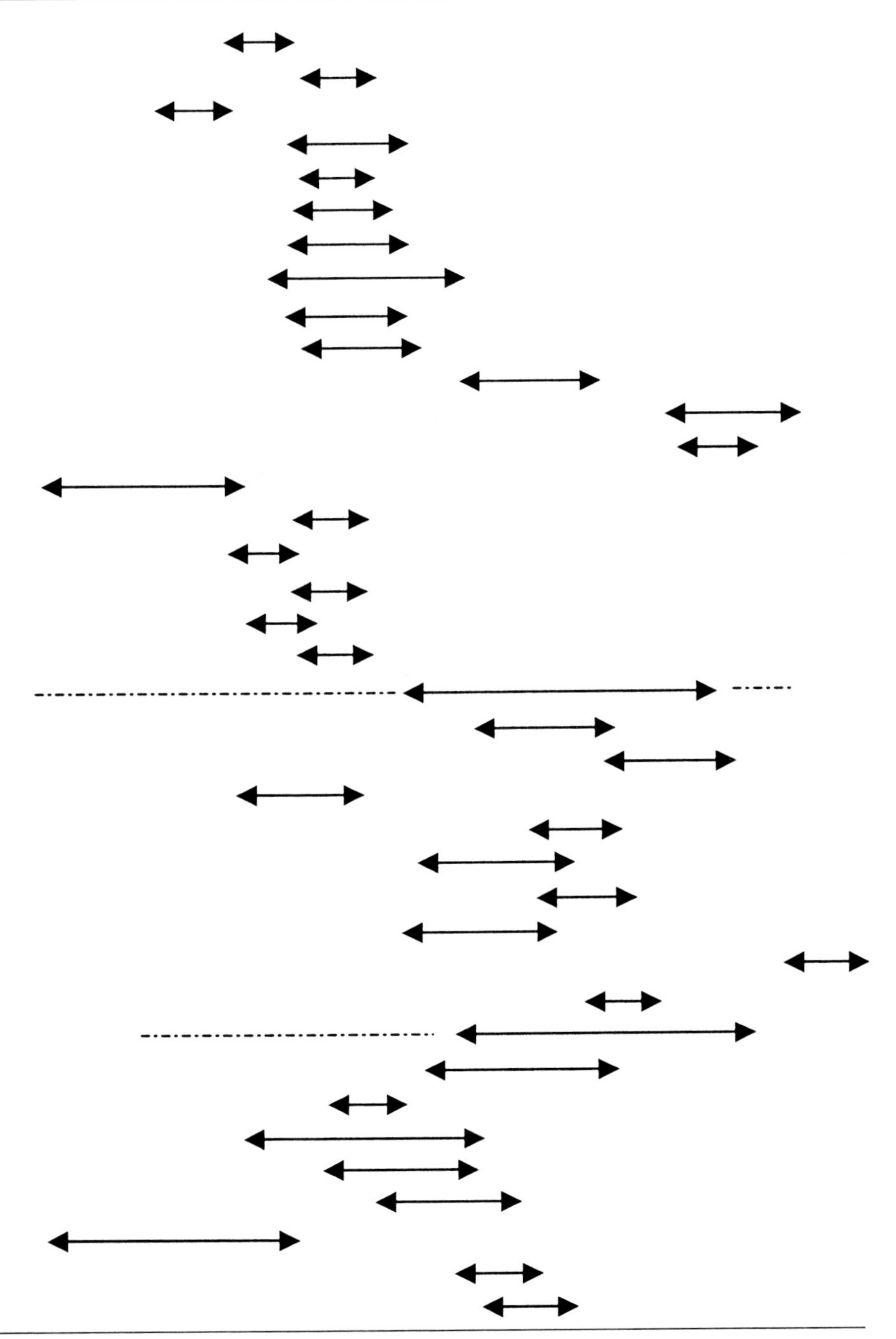

continued

Malaxis bayardii, Bayard's adder's-mouth
Malaxis spicata, Florida adder's-mouth
Malaxis unifolia, green adder's-mouth
Mesadenus lucayanus, copper ladies'-tresses
Platanthera blephariglottis, northern white fringed orchis
Platanthera chapmanii, Chapman's fringed orchis
Platanthera ciliaris, orange fringed orchis
Platanthera conspicua, southern white fringed orchis
Platanthera cristata, orange crested orchis
Platanthera flava var. *flava*, southern tubercled orchis
Platanthera flava var. *herbiola*, northern tubercled orchis
Platanthera grandiflora, large purple fringed orchis
Platanthera integrilabia, monkey-face orchis
Platanthera lacera, green fringed orchis, ragged orchis
Platanthera leucophaea, eastern prairie fringed orchis
Platanthera peramoena, purple fringeless orchis
Platanthera psycodes, small purple fringed orchis
Platythelys querceticola, low ground orchid, jug orchid
Pogonia ophioglossoides, rose pogonia
Ponthieva racemosa, shadow-witch
Pteroglossaspis ecristata, crestless plume orchid
Sacoila lanceolata, leafless beaked orchid
Spiranthes brevilabris, short-lipped ladies'-tresses
Spiranthes cernua, nodding ladies'-tresses
Spiranthes eatonii, Eaton's ladies'-tresses
Spiranthes floridana, Florida ladies'-tresses
Spiranthes lacera var. *gracilis*, southern slender ladies'-tresses
Spiranthes laciniata, lace-lipped ladies'-tresses
Spiranthes longilabris, long-lipped ladies'-tresses
Spiranthes lucida, shining ladies'-tresses
Spiranthes magnicamporum, Great Plains ladies'-tresses
Spiranthes ochroleuca, yellow ladies'-tresses
Spiranthes odorata, fragrant ladies'-tresses
Spiranthes ovalis var. *ovalis*, southern oval ladies'-tresses
Spiranthes ovalis var. *erostellata*, northern oval ladies'-tresses
Spiranthes parksii, Navasota ladies'-tresses
Spiranthes praecox, giant ladies'-tresses
Spiranthes sylvatica, woodland ladies'-tresses

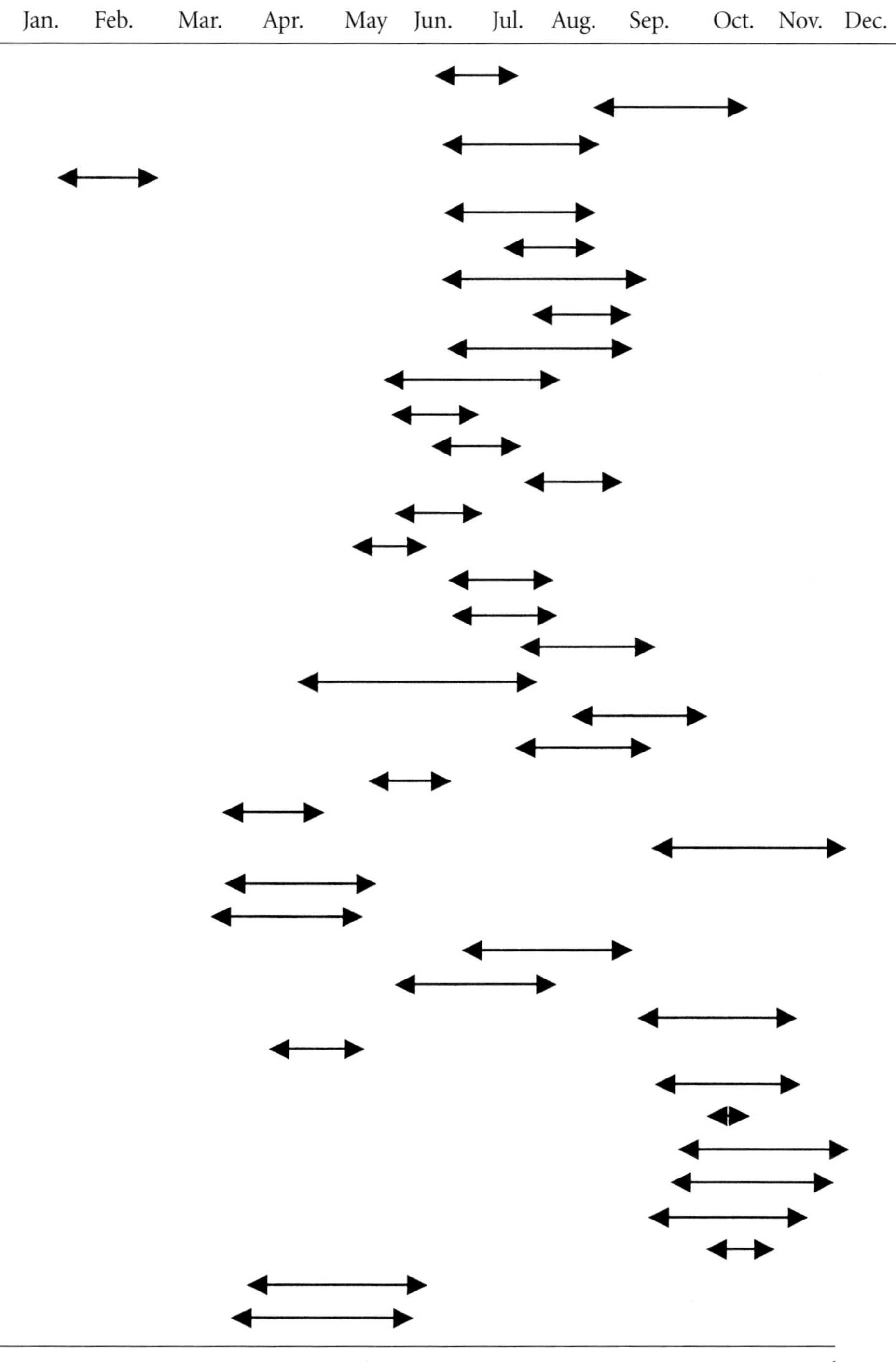

continued

Spiranthes tuberosa, little ladies'-tresses
Spiranthes vernalis, grass-leaved ladies'-tresses
Tipularia discolor, crane-fly orchis
Triphora rickettii, Rickett's noddingcaps
Triphora trianthophora, three birds orchid
Zeuxine strateumatica, lawn orchid

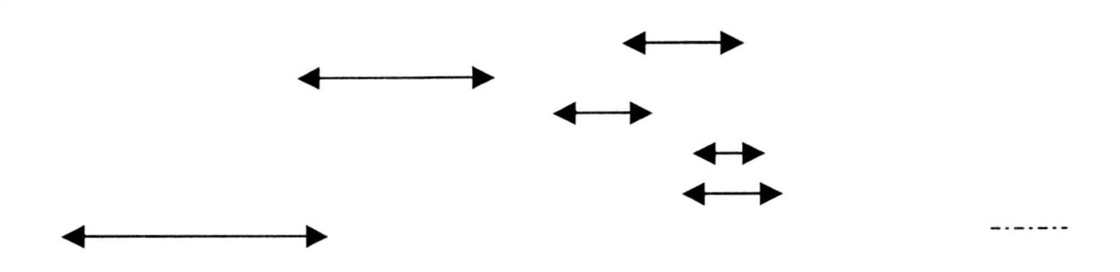
Jan.
Feb.
Mar.
Apr.
May
Jun.
Jul.
Aug.
Sep.
Oct.
Nov.
Dec.

Glossary

adventive: non-native
anterior: front or upper
anthesis: time of flowering
apomictic: fertilized within the embryo without pollination; an asexual means of reproduction
appressed: placed tightly against; opposite of divergent
approximate: the (flowers) lying close by but not overlapping
auricle: with earlike appendages
axillary: on the side
basal rosette: a circular cluster of leaves arising from the base of the plant
bract: a modified and usually reduced leaf
calcareous: derived from limestone, limy
calciphile: literally, lime lover; plant that requires lime or alkaline soil
callus: a thickened area, usually at the base of the lip
capitate: like a head; "with capitate hairs" refers to hairs with ball-like tips
carnivorous: insect eating
cauline: on the stem
chasmogamous: with fully opened, usually sexual flowers
chlorophyll: a green pigment manufactured by the plant and essential for photosynthesis
cilate: with short, slender hairs
circumneutral: a pH of about 6.5, i.e., about neutral, neither particularly acid nor alkaline
clavate: club shaped
cleistogamous: with closed flowers that are usually self-pollinating
column: the structure in an orchid that has both the anthers and the pistil
conduplicate: folded lengthwise
coralloid: coral-like
cordate: heart shaped

coriaceous: leathery
corymb: a determinate inflorescence with all of the branches the same length and the outer flowers opening first
crenulate: with a short, wavy margin
crest: a series of ridges or a group of hairs; usually yellow or a color contrasting with the lip
cyme: a determinate inflorescence with the central flowers opening first
determinate: with a specific ending, not continuing to grow indefinitely
distal: the farthest away from the center or main body; the underside
divergent: spreading or widely separated
dorsal sepal: the sepal opposite the lip; usually uppermost in most orchids
emarginate: with a short projection at the tip
endemic: native to a specific area
ephemeral: appearing for a brief time; often a few weeks each season
epiphyte, epiphytic: living in the air
erose: with an irregular margin
extant: still to be found
extirpated: no longer to be found
falcate: sickle shaped
filiform: slender and threadlike
glabrous: smooth
glaucous: with a whitish cast
habit: the way a plant grows, its pattern of branching and/or extending
habitat: where a plant grows, its environment
hemi-epiphyte: usually growing on the base of trees or on logs
hydric: a perennially wet habitat
keel: a ridge
lacerate: slashed
lateral sepals: the sepals positioned on the side of the flower
lip: the modified third petal of an orchid
lithophytic: growing on rocks
marcescent: withering but not falling off
marl: a calcareous or limy wetland
mentum: a short, rounded, thickened projection formed by the base of the petals; similar to a spur but not tapered to a point
mesic: of medium (moisture) conditions
mucronate: a short, sharp point
mycotrophic: obtaining food through mycorrhizal fungi
naturalized: a non-native species that is reproducing in its adopted habitat

neotropical: new world tropics
nominate, nominate variety: the pure species, exclusive of subspecies, variety, or form
non-resupinate: with only a single twist so the lip is at the top
oblanceolate: narrowly oblong
obovate: broadly oblong
orifice: opening
ovary: the female structure that produces the seed
panicle: a branching inflorescence similar to a raceme; the flowers stalked
pedicel: the stem of an individual flower
pedicellate flowers: those flowers held on pedicels or stalks
pedicellate ovary: typically the ovary of an orchid flower, with the ovary and flower stalk merged into one structure
peduncle: the stem of an inflorescence or aggregation of flowers
petiole: the stem portion of a leaf
plicate: soft and with many longitudinal ribs, often folded
posterior: lower or rear
pseudobulb: a swollen storage organ prominent in many epiphytic orchids and occasionally in a few genera of terrestrial orchids
puberulent: with a fine dusting of very short, soft hairs
pubescent, pubescence: downy, with short, soft hairs
putative: assumed, but not scientifically proven, in reference to hybrid parentage
raceme: an unbranched, indeterminate inflorescence with stalked flowers; branched racemes are technically panicles
reniform: kidney shaped
resupinate: twisted around so the lip is lowermost
rhizome: an elongated basal stem; typically underground in terrestrials and along the strata (tree trunks, branches, rocks, etc.) in epiphytes
rhombic: with parallel sides but tapered on both ends
rostellum: the usually beak-shaped part of the column that contains the stigmatic surface and to which the pollen adheres
saccate: shaped like a sac or pouch
saprophytic, saprophyte: obtaining food from decaying vegetable matter
scape: a leafless stem that arises from the base of the plant
secund: all to one side
segregate genus (species): a genus or species that has been separated from another larger genus or from a species
senesces: withers

sepals: the outer floral envelope
sessile: without a stem or stalk
spatulate: oblong with a narrowed base
sphagnous: an area with sphagnum moss
spike: an unbranched inflorescence with sessile or unstalked flowers
spiranthoid: a *Spiranthes* or member of a genus closely allied to *Spiranthes*
spur: a slender tubular or saclike structure usually formed at the base of the lip and often containing nectar
striate, striations: with stripes
sympatric: growing together in the same habitat
taxa (plural of taxon): particular scientific classifications or groupings such as species, subspecies, variety, form
terete: rounded
terminal: at the end
transverse: growing horizontally (as opposed to parallel) to the axis or stem
tubercle: a thickened projection
umbel: an inflorescence with pedicels arising from the end of a stem like the spokes of an umbrella
undulate: wavy
waif: a random individual occurrence
whorl: arranged in a circle around the same point on the stem
xeric: a perennially dry habitat

Selected Bibliography

Ackerman, J. D. 1995. An Orchid Flora of Puerto Rico and the Virgin Islands. *Memoirs of the New York Botanical Garden* 73.

Ames, B. 1947. *Drawings of Florida Orchids.* Cambridge, Mass.: Botanical Museum.

Ames, O. 1910. *Orchidaceae,* vol. 4, *The Genus* Habenaria *in North America.* North Easton, Mass.: Ames Botanical Laboratory.

———. 1924. *An Enumeration of the Orchids of the United States and Canada.* Boston: American Orchid Society.

———. 1937. *Zeuxine strateumatica* in Florida. *Botanical Museum Leaflet of Harvard University* 6: 37–45.

Bentley, S. L. 2000. *Native Orchids of the Southern Appalachian Mountains.* Chapel Hill: University of North Carolina Press.

Brown, P. M., and S. N. Folsom. 2002. *Wild Orchids of Florida.* Gainesville: University Press of Florida.

———. 2003. *The Wild Orchids of North America, North of Mexico.* Gainesville: University Press of Florida.

Catling, P. M. 1989. Biology of North American representatives of the subfamily Spiranthoideae, in *North American Native Terrestrial Orchid Propagation and Production.* Chadds Ford, Penn.: Brandywine Conservancy.

Correll, D. S. 1950. *Native Orchids of North America.* Waltham, Mass.: Chronica Botanica.

———. 1937. The Orchids of North Carolina. *Journal of the Elisha Mitchell Scientific Society* 53: 139–52.

Cribb, P. 1997. *The Genus* Cypripedium. Portland, Ore.: Timber Press.

Dueck, L. 2003. *Wild Orchids in South Carolina: The Story.* Aiken, S.C.: Savannah River Ecology Laboratory.

Ettman, J. K., and D. R. McAdoo. 1979. *An Annotated Catalogue and Distribution Account of the Kentucky Orchidaceae.* Louisville: Kentucky Society of Natural History Charitable Trust.

Flora of North America Editorial Committee, eds. 1993. *Flora of North America, North of Mexico,* vol. 26. New York: Oxford University Press.

Fowler, J. Forthcoming 2005. *Wild Orchids of South Carolina: A Popular Natural History.* Columbia: University of South Carolina Press.

Garay, L. A. 1982. A generic revision of the Spiranthinae. *Botanical Museum Leaflet of Harvard University* 28(4): 277–425.

Gibson, W. H. 1905. *Our Native Orchids.* New York: Doubleday, Page.

Goldman, D. H., and S. L. Orzell. 2001. Morphological, geographical, and ecological reevaluation of *Calopogon multiflorus* (Orchidaceae). *Lindleyana* 16(1): 237–51.

Gupton, O. W., and F. C. Swope. 1986. *Wild Orchids of the Middle Atlantic States.* Knoxville: University of Tennessee Press.

International Code of Botanical Nomenclature (St. Louis Code). Prepared and edited by W. Greuter, J. McNeill, F. R. Barrie, H.-M. Burdet, V. DeMoulin, T. S. Filgueiras, D. H. Nicolson, P. C. Silva, J. E. Skog, P. Trehane, N. J. Turland, and D. L. Hawksworth. 2000. *Regnum Vegetabile* 138. Online at: *http://www.bgbm.org/iapt/nomenclature/code/SaintLouis/0000St.Luistitle.htm* (accessed November 2003)

Kartesz, J. T. 1994. *A Synonymized Checklist of the Vascular Flora of United States, Canada and Greenland,* 2nd ed. Chapel Hill: North Carolina Botanical Garden.

Kartesz, J. T., and C. A. Meacham. 1999. *Synthesis of the North American Flora,* ver. 1.0. Chapel Hill: North Carolina Botanical Garden. Compact disk.

Keenan, P. E. 1999. *Wild Orchids Across North America.* Portland, Ore.: Timber Press.

Liggio, J., and A. O. Liggio. 1999. *Wild Orchids of Texas.* Austin: University of Texas Press.

Luer, C. A. 1972. *The Native Orchids of Florida.* Bronx: New York Botanical Garden.

———. 1975. *The Native Orchids of the United States and Canada excluding Florida.* Bronx: New York Botanical Garden.

Morris, M. W. 1989. *Spiranthes* in Mississippi. *Selbyana* 11: 39–48.

Morris, F., and E. Eames. 1929. *Our Wild Orchids.* New York: Charles Scribner's Sons.

Niles, G. G. 1904. *Bog Trotting for Orchids.* New York: G. P. Putnam's Sons.

Peacock, H. Big Thicket. *Texas Highways* 27(10): 14–19.

Peacock, H., and A. Garner. 1989. Orchids of the Big Thicket. *Texas Highways* 36(2): 2–9.

Petrie, W. 1981. *Guide to the Orchids of North America.* Blaine, Wash.: Hancock House.

Pridgeon, A. M., and L. E. Urbatsch. 1977. Contribution to the flora of Louisiana II: Distribution and identification of the Orchidaceae. *Castanea* 42: 293–304.

Sheviak, C. J. 1982. *Biosystematic Study of the* Spiranthes cernua *Complex.* Bulletin 448. Albany: New York State Museum.

Slaughter, C. R. 1993. *Wild Orchids of Arkansas.* Morrilton, Ark.: Privately published.

Small, J. K. 1933. *Manual of the Southeastern Flora.* New York: self-published.

Solymosy, S. L. 1963. "Orchids of Louisiana." *Bulletin, Louisiana Society for Horticultural Research* 2(3): 140–52.

Williams, J. G., A. E. Williams, and N. Arlott. 1983. *A Field Guide to Orchids of North America.* New York: Universe Books.

Wunderlin, R. P. 1998. *A Guide to the Vascular Plants of Florida.* Gainesville: University Press of Florida.

Wunderlin, R. P., and B. F. Hansen. 2000. *Atlas of Florida Vascular Plants* Web site. Tampa: Institute for Systematic Botany, University of South Florida. *http://www.plantatlas.usf.edu* (accessed November 2003)

Photo Credits

All photographs were taken by Paul Martin Brown, except for the following, which were generously loaned by those credited:

Joel DeAngelis, *Pteroglossaspis ecristata* forma *flava*
Jim Fowler, *Pteroglossaspis ecristata* forma *purpurea*
Jack Price, *Cypripedium kentuckiense* forma *pricei*
Bill Summers, *Cypripedium kentuckiense* forma *summersii*

Index

Primary entries for taxa are in **bold**. Page numbers for species descriptions are in **bold** and page numbers for photographs are in *italics*.

Related-interest titles available from University Press of Florida

The Wild Orchids of North America, North of Mexico
Paul Martin Brown and Stan Folsom

Wild Orchids of Florida
With References to the Atlantic and Gulf Coastal Plains
Paul Martin Brown and Stan Folsom

Gardening with Carnivores
Sarracenia Pitcher Plants in Cultivation and in the Wild
Nick Romanowski

Journeys Through Paradise
Pioneering Naturalists of the Southeast
Gail Fishman

The Wild East
A Biography of the Great Smoky Mountains
Margaret L. Brown

The Everglades
An Environmental History
David McCally

The Hiking Trails of Florida's National Forests, Parks, and Preserves
Johnny Molloy

Florida's Paved Bike Trails
An Eco-Tour Guide
Jeff Kunerth and Gretchen Kunerth

A Paddler's Guide to the Sunshine State
Sandy Huff

Paul Martin Brown is a research associate at the University of Florida Herbarium at the Florida Museum of Natural History in Gainesville. He is the founder of the North American Native Orchid Alliance and editor of the *North American Native Orchid Journal*. Brown and his partner Stan Folsom published *Wild Orchids of the Northeastern United States* (1997), *Wild Orchids of Florida* (UPF, 2002), and *The Wild Orchids of North America, North of Mexico* (UPF, 2003). This is their fifth volume on the wild orchids of North America.

Stan Folsom is a retired art teacher and botanical illustrator. His primary medium is watercolor; his work is represented in several permanent collections including that of the Federal Reserve Bank of Boston.